T0140863

Simon Beier

New Results on Semilinear Sets and Variants of Jumping Finite Automata

Gießener Dissertation

Fachbereich Mathematik und Informatik, Physik, Geographie

Justus-Liebig-Universität Gießen 2020

Erster Gutachter: Prof. Dr. Markus Holzer

Zweiter Gutachter: Prof. Dr. Martin Kutrib

Termin der Disputation: 21.10.2020

Bibliografische Information der Deutschen Nationalbibliothek

Die Deutsche Nationalbibliothek verzeichnet diese Publikation in der
Deutschen Nationalbibliografie; detaillierte bibliografische Daten sind
im Internet über http://dnb.d-nb.de abrufbar.

©Copyright Logos Verlag Berlin GmbH 2020

Alle Rechte vorbehalten.

ISBN 978-3-8325-5210-7

Logos Verlag Berlin GmbH

Georg-Knorr-Str. 4, Gebäude 10

D-12681 Berlin

Tel.: +49 (0)30 42 85 10 90

Fax: +49 (0)30 42 85 10 92

https://www.logos-verlag.de

What we know is a drop, what we don't know is an ocean.

Isaac Newton

Abstract

The first main part of this thesis is about semilinear sets. The descriptional complexity of operations on semilinear sets is investigated. We introduce semirecognizable subsets of monoids. A subset of a monoid is called (strongly) semirecognizable if the projection of the subset to its syntactic monoid is finite (a singleton). Thus, semirecognizable sets are a generalization of recognizable sets. The semirecognizable subsets of the free commutative monoid $\mathbb{N}^k$, which are all finite unions of strongly semirecognizable linear sets, are studied. We give characterizations in terms of rational cones when a linear set is semirecognizable and when a semirecognizable subset of $\mathbb{N}^k$ is a finite union of semirecognizable linear sets with linearly independent periods. Using the latter characterization, we show that every semirecognizable subset of $\mathbb{N}^k$ is a finite union of semirecognizable linear sets with linearly independent periods if and only if $k \leq 3$. We give characterizations in terms of linear algebra when Parikh-preimages of semirecognizable subsets of $\mathbb{N}^k$ are accepted by different kinds of pushdown automata.

The second main part of the present thesis is about variants of jumping finite automata. Jumping finite automata are defined like ordinary finite automata; however a jumping finite automaton nondeterministically jumps to an arbitrary symbol of the input string in each step of the computation and reads and consumes this symbol. The accepted languages are the permutation closed semilinear languages. We investigate the operational state complexity of jumping finite automata. A right one-way jumping finite automaton is defined like a DFA with a partial transition function. If such a device can read the current input symbol in the current internal state, the symbol is read and consumed. If the automaton cannot read the current input symbol, the read head jumps over this symbol and continues the computation with the next input symbol. When the head reaches the end of the input string, it jumps back to the beginning and goes on with its computation. We show that a permutation closed language is accepted by a right one-way jumping finite automaton if and only if it is semirecognizable. For a fixed input alphabet of cardinality at least two, the problem if the language accepted by a given right one-way jumping finite automaton is permutation closed, is **NL**-complete. While for arbitrary right one-way jumping finite automata it is unknown whether regularity, context-freeness, disjointness, inclusion, and equivalence are decidable, for right one-way jumping finite automata that accept a permutation closed language the word problem, emptiness, finiteness, universality, regularity, acceptance by different kinds of pushdown automata, disjointness, and inclusion are all in **L**, if the input alphabet is fixed. If we allow a right one-way jumping finite automaton to have multiple initial states, over a binary alphabet the accepted permutation closed languages are the finite unions of permutation closed semirecognizable languages. This characterization is generalized to arbitrary alphabets. We study different versions of nondeterministic right one-way jumping finite automata.

Collaborations

Some part of this thesis emerged from collaborative work and we would like to acknowledge these contributions. Contributions resulting from joint work with Markus Holzer and Martin Kutrib were presented at the *21st International Conference on Developments in Language Theory* [8] and the *15th International Conference on Automata and Formal Languages* [7]. Contributions resulting from joint work with Markus Holzer were presented at the *20th International Conference on Descriptional Complexity of Formal Systems* [3], the *22nd International Conference on Developments in Language Theory* [2], the *21st International Conference on Descriptional Complexity of Formal Systems* [4], and the *24th International Conference on Implementation and Application of Automata* [6]. Revised and extended journal versions of two of these contributions already appeared [5, 9].

Acknowledgments

I would like to thank Markus Holzer for supervising my PhD studies and giving me the valuable grade of freedom I needed. Thanks also go to Markus for proofreading this thesis and drawing all the contained pictures. I would like to thank Martin Kutrib for often encouraging me to finish my thesis in time. I thank my parents Oskar and Stefanie for their support during the time this thesis arose.

Contents

III Variants of Jumping Finite Automata 119

IV Outro 211

List of Figures

List of Tables

Part I
Preamble

1 Introduction

The introduction of Alan M. Turing's automatic machines [74], which are today called Turing machines, in 1936 can be seen as the starting point of automata theory. With his work in the spirit of David Hilbert's *Entscheidungsproblem* Turing aimed to give a model that is able to compute everything that can be computed by a machine. Indeed, the *Church-Turing Thesis* states that Turing machines have this ability. A *Turing machine* has a finite set of internal states which control its behaviour according to a table of rules. With a read and write head the machine has access to a tape of cells which is unbounded to the left and to the right. The input of a computation is given on this tape and at the end of the computation the output also stands on the tape. In each step of the computation the head of the machine can move one cell to the left or to the right. During the years a lot of variations of Turing machines were considered, like machines with a separate input, output, and working tape, with multiple working tapes, with a working tape that is bounded on one side, or *nondeterministic* machines, which can guess in each step of the computation how to continue. It was shown that all these models can compute the same things.

Turing machines can work as *language acceptors*, when some of the internal states are marked as accepting states. Then, the *formal language accepted by the machine* consists of all input words for which the computation of the machine halts in an accepting state. Formal languages are of great importance in computer science, for example all programming languages are formal languages, and in linguistics, where they model natural languages. With his *formal grammars*, Noam Chomsky gave a formalism trying to describe the English language and its grammar in the 1950s [16, 17]. His grammars have a finite set of rules which describe how to generate words of the formal language given by the grammar. Chomsky defined grammars of type 0, 1, 2, and 3, where grammars of type 3 have the strongest restrictions for the rules and those of type 0 have the weakest restrictions. The families of languages given by grammars of type 0, 1, 2, and 3 are the *recursively enumerable*, the *context-sensitive*, the *context-free*, and the *regular languages*. Turing machines can accept exactly the recursively enumerable languages [17].

Due to the Church-Turing Thesis there are no machine models that have a higher computational capacity than Turing machines. However these machines also have an important disadvantage: a lot of interesting problems concerning Turing machines, for example the halting problem, are undecidable, which means that they cannot be solved algorithmically or, in other words, by a Turing machine itself. *Rice's theorem* states that every non-trivial property of a language accepted by a Turing machine is undecidable. So, other automaton models, that can be seen as restricted Turing machines used as language acceptors, were introduced and investigated. For a *linear bounded automaton* the length of the working tape is linearly bounded by the length of the input. Their deterministic variant was proposed first by John R. Myhill [55]. Later Sige-Yuki Kuroda showed that nondeterministic linear bounded automata can accept exactly the context-sensitive languages [46]. It is still open today if nondeterministic linear bounded automata are strictly more powerful than their deterministic variants. Independently Anthony G. Oettinger [60] and Marcel-Paul Schützenberger [69] introduced *pushdown automata*. For these devices we have a one-way input mode and the working tape is restricted to be used as a LIFO ("last in first out") tape, which is called a *pushdown store*. Chomsky [18] and R. James Evey [26] proved that nondeterministic pushdown automata accept exactly the context-free languages, while Seymour Ginsburg and Sheila Greibach [31] and Leonard H. Haines [36] showed independently that deterministic pushdown automata are strictly less powerful than nondeterministic ones.

If we remove the pushdown store from pushdown automata, we get *finite automata*. Their first variant was given by Warren S. McCulloch and Walter Pitts in 1943 to model neural activity in nerve nets [51]. We describe finite automata in detail because they are of outstanding importance in automata theory and can be easily described due to their simple structure. A *deterministic finite automaton* consists of a finite input alphabet, a finite set of internal states, an initial state, some accepting states, and a transition function, which assigns to each pair of a state and an input symbol a successor state. The automaton starts in the initial state and reads the input from the left to the right. If it is in an accepting state after reading the last symbol, the input is accepted. Chomsky and George A. Miller notified that deterministic finite automata accept exactly the regular languages [19]. *Nondeterministic finite automata* were introduced by Michael O. Rabin and Dana Scott in 1959 [65]. In such a device the transition function can assign multiple successor states to each pair of a state and an input symbol. For each input symbol, the device guesses one of the successor states of the current state. An input is accepted if there is a sequence of guesses for each input symbol such that the automaton is in an accepting state after reading the last symbol. Rabin and Scott showed that for each nondeterministic finite automaton a deterministic one accepting the same language can be constructed *via* the powerset construction. Also, finite automata with multiple initial states and nondeterministic finite automata with spontaneous transitions have been considered. A *spontaneous transition* assigns a successor state to a state without reading an input symbol. None of these models can accept more than the regular languages.

Stephen C. Kleene proposed *regular expressions*, giving a formalism similar to arithmetic expressions that describes the family of regular languages [44]. Myhill [56] and Anil Nerode [59] proved that an arbitrary language is regular if and only if an equivalence relation induced by the language, today called the Myhill-Nerode relation, on the set of all words over the given alphabet is of finite index. Two words v and w over the alphabet are in *Myhill-Nerode relation* if the following condition holds: for all words x over the alphabet, the concatenation vx is in the language in question if and only if the concatenation wx is in the language. For a regular language, the index of the Myhill-Nerode relation equals the minimal number of states in a deterministic finite automaton accepting the language. Another equivalence relation induced by a language on the set of all words over the given alphabet, or more general an equivalence relation induced by a subset of a monoid on the monoid, is the syntactic congruence that was introduced in a paper by

Rabin and Scott and was credited to Myhill [65]. Two elements v and w of the monoid are in *syntactic congruence* if the following condition holds: for all elements x and y of the monoid, the product yvx is in the subset in question if and only if the product ywx is in the subset. The equivalence classes of the syntactic congruence build the *syntactic monoid* of the subset. Having this definition, we call the subset *recognizable* if its syntactic monoid is finite. The recognizable subsets of the *free monoid over a given alphabet*, which consists of all words over this alphabet, are the regular languages. For regular languages, given as finite automata, grammars of type 3, or regular expressions, all standard problems from formal language theory, like the word problem, emptiness, finiteness, universality, disjointness, inclusion, and equivalence, are decidable. Another exceptional feature of regular languages is that this language family is closed under all standard language operations, like union, intersection, complementation, concatenation, Kleene star, reversal, homomorphism, inverse homomorphism, and substitution.

An influential contribution to the theory of regular and context-free languages was made by Rohit J. Parikh in 1961 [61, 62]. He investigated a function, today called the *Parikh-mapping*, that maps a word over a given ordered alphabet of cardinality k to a vector in $\mathbb{N}^k$, where $\mathbb{N}$ refers to the set of non-negative integers, such that each component of the vector corresponds to the number of occurrences of one of the alphabet symbols in the word. The *Parikh-image* of a language over the given alphabet is a subset of $\mathbb{N}^k$, namely the image of the language under the Parikh-mapping. Now, Parikh defined semilinear subsets of $\mathbb{N}^k$ as follows in terms of linear algebra. A *linear* subset of $\mathbb{N}^k$ is a set of the form

$$\left\{ \vec{c} + \sum_{\vec{p} \in P} \lambda_{\vec{p}} \cdot \vec{p} \;\middle|\; \forall \vec{p} \in P : \lambda_{\vec{p}} \in \mathbb{N} \right\}$$

for a so called *constant vector* $\vec{c} \in \mathbb{N}^k$ and a finite set P of so called *periods*, which are vectors in $\mathbb{N}^k$. A *semilinear* subset of $\mathbb{N}^k$ is a finite union of linear subsets. It is not hard to see that for each semilinear set one can construct a regular expression describing a language whose Parikh-image is the semilinear set. So, each semilinear set is the Parikh-image of a regular language. On the other hand, Parikh showed that for each context-free language a representation of the Parikh-image as a semilinear set can be computed, which implies that for each context-free language a regular language having the same Parikh-image can be computed. Because the family of semilinear sets equals the set of Parikh-images of regular languages and the set of Parikh-images of context-free languages, semilinear sets are of great importance in formal language theory.

Clearly, the family of semilinear sets is closed under the operations of union, direct product, sum of two such sets, and under building the image under a linear map which maps vectors from $\mathbb{N}^k$ to vectors from $\mathbb{N}^k$. Ginsburg and Edwin H. Spanier showed that the family of semilinear sets is also closed under the operations of intersection, complementation, and building the preimage under a linear map which maps vectors from $\mathbb{N}^k$ to vectors from $\mathbb{N}^k$ [32]. Semilinear sets are related to *Presburger arithmetic*, which is the predicate logic of the integers with addition and was introduced by Mojżesz Presburger in 1929 [64]. He proved that all *Presburger sentences*, *i.e.* Presburger formulas without free variables, are decidable [38, 64]. It was shown by Ginsburg and Spanier that the semilinear sets are those subsets of $\mathbb{N}^k$ which can be described by a Presburger formula [34]. Ginsburg and Spanier also proved that each semilinear set is a finite union of linear sets with linearly independent periods [32]. This is a discrete analogue of the conical version of Carathéodory's Theorem from convex geometry, which states that for each subset S of the Euclidean space the conical hull of S equals the union of the conical hulls of the linearly independent subsets of S. Later, Samuel Eilenberg and Schützenberger [24] and Ryuichi Ito [43] independently showed that each semilinear set even is a finite union of disjoint linear sets with linearly independent periods.

In the last one and a half decades there was a lot of interest in variants of finite automata which do not read their input from left to right, but in a discontinuous way. Suna Bensch *et al.* studied *input-reversal* and *input-revolving finite automata* [10, 11, 12, 13]. In each step of the computation such a device can either read and consume the current input symbol like an ordinary finite automaton or it can change the order of the letters in the input string depending on the current internal state of the device and on the currently read input symbol, in the latter case the current input symbol is

not consumed. While input reversal finite automata can reverse the whole
input string, input-revolving finite automata can do three types of revolving
transitions: a *left-revolving transition* moves the last input symbol to the
beginning of the input string, a *right-revolving transition* moves the first
input symbol to the end of the input string, and a *circular interchanging
transition* swaps the first and the last symbol of the input string. László
Kovács, Benedek Nagy, and Friedrich Otto investigated *finite-state acceptors with translucent letters* [57, 58]. For these devices in each step of the
computation some letters of the input alphabet, depending on the current
internal state, are translucent, which means that the automaton does not
see them and reads and consumes the first letter in the input string which
is not translucent. The input is accepted if there is a computation which
leads to a situation where all remaining input symbols are translucent and
the automaton is in an accepting state.

Alexander Meduna and Petr Zemek introduced *jumping finite automata*
in 2012 [53, 54]. Such an automaton is defined like an ordinary finite automaton, however the way it processes the input is different. A jumping
finite automaton nondeterministically jumps to an arbitrary symbol of the
input string in each step of the computation and reads and consumes this
symbol. Here, the input is accepted in case that there is a computation
that leads to a situation where the whole input is read and the device is
in an accepting state. If we interpret a finite automaton as a jumping finite automaton, the accepted language is the permutation closure of the
language accepted by the original finite automaton, where the *permutation closure* of a language consists of all permutations of all words from
the language. So, the languages accepted by jumping finite automata are
the permutation closures of regular languages. It is not hard to see that
these are the permutation closed languages with a semilinear Parikh-image,
where a language is called *permutation closed* if it equals its permutation
closure. Hence, there is a one-to-one correspondence between languages
accepted by jumping finite automata and semilinear sets. It follows that
the word problem, emptiness, finiteness, universality, disjointness, inclusion,
and equivalence are all decidable for jumping finite automata. Regularity is
decidable for jumping finite automata as well because Ginsburg and Spanier
showed that one can decide if the permutation closure of a regular language

is regular [33].

Jumping finite automata were also studied by Henning Fernau *et al.* [28, 29]. The family of languages accepted by jumping finite automata is closed under union and is not closed under the operations of intersection with regular languages, concatenation, Kleene star, homomorphism, and substitution, as shown by Meduna and Zemek. They also claimed that this family is closed under intersection, complementation, and inverse homomorphism. Fernau *et al.* pointed out that the proofs for intersection and complementation are do not work, but that the family of languages accepted by jumping finite automata is nevertheless closed under intersection and complementation because semilinear sets are closed under these operations. We show that the proof for inverse homomorphism by Meduna and Zemek also does not work. However, the family of languages accepted by jumping finite automata is closed under inverse homomorphism because semilinear sets are.

Right one-way jumping finite automata, a variant of jumping finite automata, were investigated by Hiroyuki Chigahara *et al.* in 2016 [14]. *Right one-way jumping finite automata* are defined like ordinary deterministic finite automata with a partial transition function, but process the input differently than finite automata and jumping finite automata. If a right one-way jumping finite automaton can read the current input symbol in the current internal state, the symbol is read and consumed. If the device cannot read the current input symbol, the read head jumps over this symbol and continues the computation with the next input symbol. When the head reaches the end of the input string, it jumps back to the beginning of the string and goes on with its computation. So, one can imagine the input string as a cycle, where the head of the automaton starts at some given position, which corresponds to the beginning of the input string. However, the device does not remember this position after starting the computation. The input is accepted if this computation, which is deterministic, leads to a situation where the whole input is read and the device is in an accepting state.

The computation of a right one-way jumping finite automaton can also be imagined differently: if the symbol at the beginning of the input string cannot be read, this symbol is moved to the end of the input string. This way a right one-way jumping finite automaton can be interpreted as a deterministic right-revolving finite automaton as defined by Bensch *et al.* [11]. Thus, every language accepted by a right one-way jumping finite automaton is also accepted by a deterministic right-revolving finite automaton. The difference between these two automata models is that a right-revolving finite automaton is also allowed to change its internal state without consuming the current input symbol. Chigahara *et al.* gave a sketch of a proof that there is a language accepted by a deterministic right-revolving finite automaton, but not by a right one-way jumping finite automaton. Clearly, there is also a similarity between right one-way jumping finite automata and deterministic finite-state acceptors with translucent letters as defined by Nagy and Otto [58]. Here, the difference is that the head of a finite-state acceptor with translucent letters jumps back to the beginning of the input string after each step of the computation and that a finite-state acceptor with translucent letters can also accept the input in a situation where not all of the input symbols are read.

For each deterministic finite automaton with a partial transition function, the language accepted by the automaton interpreted as a right one-way jumping finite automaton is sandwiched between the language accepted by the original finite automaton and its permutation closure, which equals the language accepted by the automaton interpreted as a jumping finite automaton. Hence, these three languages have the same Parikh-image, which is a semilinear set. If the transition function of the automaton is total, the language accepted by the automaton interpreted as a right one-way jumping finite automaton equals the language accepted by the original finite automaton. That is why each regular language is accepted by a right one-way jumping finite automaton. Chigahara *et al.* showed that the family of languages accepted by right one-way jumping finite automata is incomparable to the context-free languages.

This thesis is divided into four parts. The first one, this preamble, continues with a chapter that explains the basic notions used in the thesis. In the second part new results about semilinear sets are presented in two chapters: The third chapter treats the descriptional complexity of operations on semilinear sets. We give upper bounds for the parameters of the resulting semilinear set when one of the operations intersection, complementation, or inverse homomorphism is applied to semilinear sets. To do so, we analyse the original proofs of Ginsburg and Spanier that semilinear sets are closed under these operations [32]. The fourth chapter deals with semirecognizable subsets of $\mathbb{N}^k$. Clearly, a permutation closed language is regular if and only if the Parikh-image of the language, as a subset of the monoid $\mathbb{N}^k$ with addition, is recognizable. The recognizable subsets of $\mathbb{N}^k$ are well understood because *Mezei's Theorem* states that these are the finite unions of direct products of semilinear subsets of $\mathbb{N}$ [25]. Our focus lies on a generalization of recognizable subsets of a monoid, namely those subsets for which the projection of the subset to the syntactic monoid is finite; we call them *semirecognizable subsets*. If the projection of a subset to the syntactic monoid is a singleton or the empty set, the subset is called *strongly semirecognizable*. The semirecognizable subsets of $\mathbb{N}^k$, which are all finite unions of strongly semirecognizable linear sets, are investigated in this section. We characterize when a linear subset of $\mathbb{N}^k$ is semirecognizable and show a connection to rational cones. Then, *lattices* are introduced as special strongly semirecognizable sets that extend "the pattern" of a linear set to the whole $\mathbb{N}^k$. Each strongly semirecognizable subset of $\mathbb{N}^k$ equals a lattice if we only consider vectors where all components are "large enough". We study Carathéodory-like decompositions of lattices and arbitrary (strongly) semirecognizable subsets of $\mathbb{N}^k$. More precisely, we prove a characterization when a lattice or an arbitrary (strongly) semirecognizable subset of $\mathbb{N}^k$ is a finite union of (strongly) semirecognizable linear sets with linearly independent periods; again there is a connection to rational cones. A result is given that states in terms of lattices when an arbitrary subset of $\mathbb{N}^k$ is a finite union of semirecognizable subsets of $\mathbb{N}^k$. Parikh-preimages of semirecognizable subsets of $\mathbb{N}^k$ are considered. These are the permutation closed semirecognizable languages, which build a proper subfamily of the permutation closed languages with a semilinear Parikh-image. Michel Latteux proved that over a binary alphabet each permutation closed language with a semilinear Parikh-image is context free [48]. Michel Rigo gave an easier proof of this considerable result using context-free grammars and factorization of words [66]. We give another even more elementary proof of the same result using counter automata. Several characterizations in terms of linear algebra, when permutation closed semirecognizable languages are accepted by different types of pushdown automata, are stated.

The third part of the thesis gives new results about variants of jumping finite automata: In the fifth chapter the operational state complexity of jumping finite automata is investigated. We provide upper bounds for the number of states of a jumping finite automaton accepting the resulting language when one of the operations intersection, complementation, or inverse homomorphism is applied to languages accepted by jumping finite automata. Also the operation of intersection of a language accepted by a jumping finite automaton and a language accepted by an ordinary finite automaton is treated in the case where the resulting language is accepted by a jumping finite automaton. To get our bounds we use the results about the descriptional complexity of operations on semilinear sets from Chapter 3. The sixth chapter deals with properties of right one-way jumping finite automata. We show that a permutation closed language is accepted by a right one-way jumping finite automaton if and only if it is semirecognizable. Closure properties of the family of languages accepted by right one-way jumping finite automata are given. We prove characterizations of languages accepted by right one-way jumping finite automata when the languages are given as a concatenation of a language and a word or as a concatenation of two languages over disjoint alphabets. Decidability and complexity of problems involving right one-way jumping finite automata are studied in Chapter 7. While for this model the word problem, emptiness, finiteness, and universality are easily decidable, it is not known if regularity, context-freeness, disjointness, inclusion, and equivalence are decidable. The problem whether a given word is the prefix of a word accepted by a given right one-way jumping finite automaton and similar problems can be decided by a Turing machine that only needs polynomial space. It can be decided by a nondeterministic Turing machine that only needs logarithmic space whether the language accepted by a given right one-way jumping finite automaton over a fixed alphabet is permutation closed. For a right one-way jumping finite automaton that accepts a permutation closed language over a fixed alphabet, the word problem, emptiness, finiteness, universality, regularity, acceptance by different kinds of pushdown automata, disjointness, and inclusion can be decided by a deterministic Turing machine that only needs logarithmic space. *Nondeterministic right one-way jumping finite automata, i.e.,* different variants of nondeterministic finite automata interpreted as right one-way jumping finite automata, are investigated in the eighth chapter. A characterization of permutation closed languages accepted by right one-way jumping finite automata with multiple initial states is shown using results from Chapter 4. This general characterization becomes stronger for binary alphabets: there, a permutation closed language is accepted by a right one-way jumping finite automaton with multiple initial states if and only if it is a finite union of permutation closed semirecognizable languages. We also prove that each language accepted by a jumping finite automaton is accepted by a nondeterministic right one-way jumping finite automaton with a single initial state and without spontaneous transitions. Inclusion relations of families of languages accepted by nondeterministic right one-way jumping finite automata to families of languages accepted by right-revolving finite automata or finite-state acceptors with translucent letters are given. The fourth part of the present thesis is the outro which discusses conclusion and directions for further research.

2 Basic Notions

In this chapter the basic notions and concepts that we work with are discussed. That enables the interested reader with a background in computer science and mathematics to follow our investigations. For detailed introductions to formal languages and automata, we refer to the books *Introduction to Automata Theory, Languages, and Computation* by John E. Hopcroft and Jeffrey D. Ullman [39], *Introduction to Formal Language Theory* by Michael A. Harrison [37], and *Formal Languages* by Arto Salomaa [67]. The first section of this chapter explains the essential notions concerning monoids and recognizable subsets of monoids. A *monoid* is an algebraic structure that can be seen as an abstraction of the set of all words over a given alphabet. A *recognizable* subset of a monoid is a preimage of some set under a monoid homomorphism mapping to a finite monoid. Recognizable sets are an abstraction of regular languages. In the second section we deal with important operations on formal languages or, more general, on subsets of monoids. The third section is about Turing machines. Pushdown automata and the class of languages accepted by them, the context-free languages, are introduced in the fourth section, whereas in the fifth section finite automata are treated. The family of Parikh-images of regular languages equals the family of Parikh-images of context-free languages. The basics about this family of subsets of $\mathbb{N}^k$, the semilinear sets, are given in the sixth section. We explain jumping finite automata in the seventh section and right one-way jumping finite automata in the eighth section.

We use $\subseteq$ for inclusion and $\subset$ for proper inclusion of sets. For sets S and T the set of all maps from S to T is denoted by T^S while 2^S stands for the powerset of S. For sets S and T with $S \subseteq T$ let $\iota_{S \to T} \in T^S$ be the inclusion. For a set S and a binary relation $\sim$ on S, let $\sim^+$ be the transitive closure of $\sim$ and $\sim^*$ be the reflexive-transitive closure of $\sim$. Furthermore, for $n \geq 0$ let $\sim^n$ be the binary relation on S which is defined as follows. For $a, b \in S$ we have $a \sim^n b$ if there are $c_0, c_1, \ldots, c_n \in S$ with $c_0 = a$ and $c_n = b$ such that for all $i \in \{0, 1, \ldots, n-1\}$ it holds $c_i \sim c_{i+1}$. Let $\mathbb{N}$ be the set of nonnegative integers, $\mathbb{Z}$ be the set of integers, $\mathbb{Q}$ be the set of rational numbers, and $\mathbb{R}$ be the set of real numbers. We use the convention $\max(\emptyset) = 0$. Let $k \geq 0$. For $\vec{x} = (x_1, x_2, \ldots, x_k) \in \mathbb{N}^k$ let $||x||_\infty = \max(\{\, x_i \mid i \in \{1, 2, \ldots, k\} \,\})$ be the *maximum norm* of $\vec{x}$. For a finite $F \subseteq \mathbb{N}^k$ let $||F||_\infty = \max(\{\, ||\vec{x}||_\infty \mid \vec{x} \in F \,\})$. For $T \subseteq \{1, 2, \ldots, k\}$ with $T = \{t_1, t_2, \ldots, t_{|T|}\}$ and $t_1 < t_2 < \cdots < t_{|T|}$ define $\pi_{k,T} : \mathbb{N}^k \to \mathbb{N}^{|T|}$ as

$$\pi_{k,T}(x_1, x_2, \ldots, x_k) = (x_{t_1}, x_{t_2}, \ldots, x_{t_{|T|}}).$$

For a number $t \in \{1, 2, \ldots, k\}$ we also write $\pi_{k,t}$ instead of $\pi_{k,\{t\}}$ sometimes. Let $\overrightarrow{e_{k,1}}, \overrightarrow{e_{k,2}}, \ldots, \overrightarrow{e_{k,k}}$ be the standard basis of the k-dimensional Euclidean space. For an $S \subseteq \mathbb{R}^k$ let $\mathsf{span}(S)$ be the linear subspace of $\mathbb{R}^k$ spanned by S. An $S \subseteq \mathbb{R}^k$ is called *collinear* if there is an $\vec{x} \in \mathbb{R}^k$ such that $S \subseteq \{\, \lambda\vec{x} \mid \lambda \in \mathbb{R} \,\}$. The elements of the set $\mathbb{N}^k$ can be partially ordered by the $\leq$-relation on vectors. For vectors $\vec{x}, \vec{y} \in \mathbb{N}^k$ we write $\vec{x} \leq \vec{y}$ if all components of $\vec{x}$ are less or equal to the corresponding components of $\vec{y}$. In this way we especially can speak of *minimal elements* of subsets of $\mathbb{N}^k$. In fact, due to Dickson's Lemma every subset of $\mathbb{N}^k$ has only a finite number of minimal elements [22]. Let $S \subseteq \mathbb{N}^k$. Then, denote by $\min(S) \subseteq S$ the set of minimal elements of S concerning the $\leq$ relation on $\mathbb{N}^k$. For $\vec{x} \in \mathbb{N}^k$ let $S_{\geq \vec{x}} = \{\, \vec{s} \in S \mid \vec{x} \leq \vec{s} \,\}$. For $T \subseteq \mathbb{N}^k$ set

$$S - T = \left\{\, \vec{s} - \vec{t} \,\middle|\, \vec{s} \in S, \vec{t} \in T \,\right\} \subseteq \mathbb{Z}^k,$$

which should not be confused with the set difference $S \setminus T$.

2.1 Monoids and Recognizable Subsets

A fundamental role in formal language theory is played by monoids, so some elementary definitions and properties concerning monoids are given. A *monoid* is a triple $(M, \cdot, e)$, where M is a set, $\cdot$ is a binary operation of the form $M \times M \to M$, and e is an element of M, such that $\cdot$ is *associative*, meaning that for all $a, b, c \in M$ we have $(a \cdot b) \cdot c = a \cdot (b \cdot c)$, and e is an *identity element*, meaning that for all $a \in M$ it holds $e \cdot a = a \cdot e = a$. Because of associativity we can omit brackets. For a monoid $(M, \cdot, e)$ and $a, b \in M$, we also write ab instead of $a \cdot b$. Having a monoid $(M, \cdot, e)$ we sometimes also call the set M a monoid if it is clear which binary operation we are considering, obviously e is already determined by M and the binary operation. For a monoid $(M, \cdot, e)$ and $a \in M$ we set $a^0 = e$ and for $n \geq 1$ we set $a^n = a^{n-1}a$. A monoid $(M, \cdot, e)$ is called *commutative* if for all $a, b \in M$ we have $ab = ba$. In a commutative monoid the binary operation is often written as $+$ and the identity element as 0. For a commutative monoid $(M, +, 0)$, $a \in M$, and $n \geq 0$ we write $n \cdot a$ or just na instead of a^n. For a commutative monoid $(M, +, 0)$, $S \subseteq M$, and $n \geq 0$ we set $nS = n \cdot S = \{ na \mid a \in S \}$. For a monoid $(M, \cdot, e)$ a subset $S \subseteq M$ is called a *submonoid* of M if $e \in S$ and for all $a, b \in S$ it holds $ab \in S$, in this case we write $S \subseteq_{\mathrm{sub}} M$. For monoids $(M_1, \cdot_1, e_1)$ and $(M_2, \cdot_2, e_2)$ a map $f : M_1 \to M_2$ is called a *monoid homomorphism* if $f(e_1) = e_2$ and for all $a, b \in M_1$ we have $f(a \cdot_1 b) = f(a) \cdot_2 f(b)$. A bijective monoid homomorphism is called a *monoid isomorphism*. For $k, \ell \geq 0$ and a monoid homomorphism $h : \mathbb{N}^k \to \mathbb{N}^\ell$ let

$$||h||_{\max} = \max \left(\{ \, ||h\, (\overrightarrow{e_{k,i}})||_\infty \mid i \in \{1, 2, \ldots, k\} \, \} \right).$$

For a set S the *free monoid over* S is the monoid $(S^*, \cdot_S, \lambda_S)$, where

$$S^* = \{ \, a_1 a_2 \cdots a_n \mid n \geq 0, \, a_1, a_2, \ldots, a_n \in S \, \}$$

is the set of finite sequences of elements from S and the binary operation $\cdot_S$ is given as follows. For $n, m \geq 0$ and $a_1, a_2, \ldots, a_n, b_1, b_2, \ldots, b_m \in S$ we set

$$a_1 a_2 \cdots a_n \cdot_S b_1 b_2 \cdots b_m = a_1 a_2 \cdots a_n b_1 b_2 \cdots b_m.$$

The identity element λ_S is the empty sequence, which is often just denoted as λ. Let $\iota_S : S \to S^*$ be the injective map given by $a \mapsto a$. Then, the free monoid over S has the following universal property. For each monoid $(M, \cdot, e)$ and each map $f : S \to M$ there is a unique monoid homomorphism $f^* : S^* \to M$ such that $f^* \circ \iota_S = f$.

A nonempty finite set is called an *alphabet*. For an alphabet Σ, the elements of Σ are called *symbols*, the elements of the free monoid Σ^* are called *words*, and the subsets of Σ^* are called *languages over* Σ. Let **FIN** be the class of all finite languages. For $n \geq 0$ and $a_1, a_2, \ldots, a_n \in \Sigma$ the *reversal* w^R of the word $w = a_1 a_2 \cdots a_n$ is $a_n a_{n-1} \cdots a_1$. For $n \geq 0$ and $a_1, a_2, \ldots, a_n \in \Sigma$ the *length* $|w|$ of the word $w = a_1 a_2 \cdots a_n$ is n. For an $a \in \Sigma$ and a $w \in \Sigma^*$ the number of occurrences of a in w is denoted by $|w|_a$. For two alphabets Σ and Γ and a monoid homomorphism $h : \Sigma^* \to \Gamma^*$ let $|h| = \max\left(\{\, |h(a)|_b \mid a \in \Sigma, b \in \Gamma \,\}\right)$ be the *absolute value* of h. Let $v, w \in \Sigma^*$. We say that v is a *prefix* of w if there is an $x \in \Sigma^*$ with $w = vx$. We say that v is a *suffix* of w if there is an $x \in \Sigma^*$ with $w = xv$. We say that v is a *factor* of w if there are $x, y \in \Sigma^*$ with $w = xvy$. We say that v is a *sub-word* of w if there are an $n \geq 0$ and words $x_1, x_2, \ldots, x_n, y_1, y_2, \ldots, y_{n+1} \in \Sigma^*$ with $v = x_1 x_2 \cdots x_n$ and $w = y_1 x_1 y_2 x_2 \cdots y_n x_n y_{n+1}$. A language $L \subseteq \Sigma^*$ is called *prefix-free* if there are no $v, w \in L$ such that $v \neq w$ and v is a prefix of w.

For a set S the *free commutative monoid over* S is defined to be the commutative monoid $(\mathrm{Map}_{\mathrm{fin}}(S, \mathbb{N}), +_S, 0_S)$, where

$$\mathrm{Map}_{\mathrm{fin}}(S, \mathbb{N}) = \left\{ f \in \mathbb{N}^S \,\middle|\, |\{\, a \in S \mid f(a) > 0 \,\}| < \infty \right\},$$

the symbol $+_S$ denotes the pointwise addition of maps, and $0_S : S \to \mathbb{N}$ is the constant 0-function. Let $\kappa_S : S \to \mathrm{Map}_{\mathrm{fin}}(S, \mathbb{N})$ be the injective map which is defined through

$$a \mapsto \left(b \mapsto \begin{cases} 1 & \text{if } b = a \\ 0 & \text{otherwise} \end{cases} \right).$$

Now, the following universal property is fulfilled by the free commutative monoid over S. For each commutative monoid $(M, +, 0)$ and each map $f : S \to M$ there is a unique monoid homomorphism $g : \mathrm{Map}_{\mathrm{fin}}(S, \mathbb{N}) \to M$ such that $g \circ \kappa_S = f$.

For an alphabet Σ, an order on Σ gives us a canonical isomorphism between the two monoids $(\mathrm{Map}_{\mathrm{fin}}(\Sigma, \mathbb{N}), +_\Sigma, 0_\Sigma)$ and $\left(\mathbb{N}^{|\Sigma|}, +, \vec{0} \right)$, where $+$ is the ordinary vector addition and $\vec{0}$ is the vector whose components are all 0.

For an ordered alphabet $\Sigma = \{a_1, a_2, \ldots, a_{|\Sigma|}\}$ the universal property of the free monoid over Σ gives us a unique monoid homomorphism $\psi_\Sigma : \Sigma^* \to \mathbb{N}^{|\Sigma|}$ such that for all $i \in \{1, 2, \ldots, |\Sigma|\}$ we have that $\psi_\Sigma(a_i) = \overrightarrow{e_{|\Sigma|,i}}$. This map ψ_Σ is called the *Parikh-mapping*. For each word $w \in \Sigma^*$ the vector $\psi_\Sigma(w)$ is called the *Parikh-vector* of w and for each language $L \subseteq \Sigma^*$ the image $\psi_\Sigma(L)$ is called the *Parikh-image* of L. Two languages over Σ^* are called *Parikh-equivalent* if they have the same Parikh-image. For a language $L \subseteq \Sigma^*$ the *permutation closure* of L is $\mathsf{perm}(L) = \psi_\Sigma^{-1}(\psi_\Sigma(L))$. A language $L \subseteq \Sigma^*$ is called *permutation closed* if $L = \mathsf{perm}(L)$. If it is clear which ordered alphabet we are talking about, we sometimes just write ψ instead of ψ_Σ. For a class **FAM** of languages, let **pFAM** be the class of all permutation closed languages in **FAM**. If we have two formal descriptions of the same language, the two descriptions are called *equivalent*. If we do not demand that the two descriptions describe the same language but that the two described languages are Parikh-equivalent, the two descriptions are called *Parikh-equivalent*.

For a monoid $(M, \cdot, e)$, a subset $S \subseteq M$ is called *recognizable* if there exist a finite monoid $(M_0, \cdot_0, e_0)$ and a monoid homomorphism $f : M \to M_0$ such that $S = f^{-1}(f(S))$. Thus, for each $a \in M$ the value $f(a)$ tells us if $a \in S$.

Example 2.1. Consider the language $L = \{\, (ab)^n \mid n \geq 0 \,\} \subset \{a, b\}^*$. Let $M_0 = \{e_0, s\} \cup \{a, b\}^2$ and $\cdot_0$ be the binary operation on M_0 defined as follows. For all $c \in M_0$ set $e_0 \cdot_0 c = c \cdot_0 e_0 = c$ and $s \cdot_0 c = c \cdot_0 s = s$. For $(c_1, c_2), (d_1, d_2) \in \{a, b\}^2$ set

$$(c_1, c_2) \cdot_0 (d_1, d_2) = \begin{cases} s & \text{if } c_2 = d_1, \\ (c_1, d_2) & \text{otherwise.} \end{cases}$$

Then, $(M_0, \cdot_0, e_0)$ is a monoid. Let the map $f : \{a, b\}^* \to M_0$ be defined as follows. We set $f(\lambda) = e_0$ and for all words $w \in \{a, b\}^*$ for which aa or bb is a factor, let $f(w) = s$. For all $n \geq 1$ and $c_1, c_2, \ldots, c_n \in \{a, b\}$ such that for all $i \in \{1, 2, \ldots, n-1\}$ it holds $c_i \neq c_{i+1}$, we set $f(c_1 c_2 \cdots c_n) = (c_1, c_n)$. The function f is a monoid homomorphism. We have $f^{-1}(f(L)) = f^{-1}(\{e_0, (a, b)\}) = L$. Hence, the language L is recognizable.

Recognizability of a subset of a monoid can be characterized *via* the syntactic monoid of the subset, as we demonstrate now. For a monoid $(M, \cdot, e)$ and a subset $S \subseteq M$ the *syntactic congruence* $\sim_S$ is an equivalence relation on M, which is defined as follows. For $a, b \in M$ we have $a \sim_S b$ if for all $c, d \in M$ the condition $cad \in S$ is equivalent to the condition $cbd \in S$. Let $a, b, c, d \in M$ with $a \sim_S b$ and $c \sim_S d$. For all $p, q \in M$ we have

$$pacq \in S \Leftrightarrow padq \in S \Leftrightarrow pbdq \in S.$$

It follows $ac \sim_S bd$. So, $M/\sim_S$ becomes a monoid, which is called the *syntactic monoid of S*. The canonical projection $\pi_{\sim_S} : M \to M/\sim_S$ is a monoid homomorphism. We have

$$\pi_{\sim_S}^{-1}\left(\pi_{\sim_S}(S)\right) = \pi_{\sim_S}^{-1}\left(\{\,[a]_{\sim_S} \mid a \in S\,\}\right) = S$$

because for all $a \in S$ we have $eae \in S$, while for all $b \in M \setminus S$ it holds $ebe \notin S$. The syntactic monoid has the following property.

Proposition 2.2. *Let $(M, \cdot, e)$ and $(M_0, \cdot_0, e_0)$ be monoids, $S \subseteq M$ be a subset, and $f : M \to M_0$ be a monoid homomorphism with the property that $S = f^{-1}(f(S))$. Then, there exists a surjective monoid homomorphism $g : f(M) \to M/\sim_S$ with $g \circ f = \pi_{\sim_S}$.*

Proof. Consider $a, b \in M$ with $f(a) = f(b)$. For all $c, d \in M$ it holds

$$cad \in S \Leftrightarrow f(cbd) = f(c)f(b)f(d) = f(c)f(a)f(d) = f(cad) \in f(S)$$
$$\Leftrightarrow cbd \in S,$$

which gives us $a \sim_S b$. Hence the map $g : f(M) \to M/\sim_S$, $f(a) \mapsto [a]_{\sim_S}$ is well-defined. Also, it clearly is a surjective monoid homomorphism fulfilling $g \circ f = \pi_{\sim_S}$. $\quad\square$

As a direct consequence we get:

Proposition 2.3. *Consider a monoid $(M, \cdot, e)$ and a subset $S \subseteq M$. Then, S is recognizable if and only if its syntactic monoid is finite.* $\quad\square$

Example 2.4. Consider the language $L \subset \{a, b\}^*$, the monoid $(M_0, \cdot_0, e_0)$, and the monoid homomorphism $f : \{a, b\}^* \to M_0$ from Example 2.1. It is not hard to see that for all $v, w \in \{a, b\}^*$ the condition $v \sim_L w$ holds if and only if $f(v) = f(w)$. Hence, we get a monoid isomorphism of the form $\{a, b\}^*/\sim_L \to M_0$ through $[w]_{\sim_L} \mapsto f(w)$. So, the syntactic monoid of L consists of six elements.

If you are given two monoids $(M_1, \cdot_1, e_1)$ and $(M_2, \cdot_2, e_2)$, the direct product $(M_1 \times M_2, \cdot_1 \times \cdot_2, (e_1, e_2))$ is again a monoid. Mezei's Theorem characterizes the recognizable subsets of a direct product of monoids:

Theorem 2.5 ([25]). *Let $(M_1, \cdot_1, e_1)$ and $(M_2, \cdot_2, e_2)$ be monoids and $S \subseteq M_1 \times M_2$. Then, the subset S is recognizable if and only if there are an $n \geq 0$ and recognizable subsets $T_1, T_2, \ldots, T_n$ of M_1 and $U_1, U_2, \ldots, U_n$ of M_2 such that $S = \bigcup_{i=1}^{n}(T_i \times U_i)$.*

We give a proof of the following well-known fact concerning the free (commutative) monoid $\mathbb{N}$.

Proposition 2.6. *Each submonoid of $\mathbb{N}$ is recognizable.*

Proof. Let $S \subseteq \mathbb{N}$ be a submonoid with $\{0\} \subset S$. Set $m = \min(S \setminus \{0\})$ and furthermore $R = \{a \bmod m \mid a \in S\}$ and

$$n = \max\left(\{\min(\{a \in S \mid a \bmod m = r\}) \mid r \in R\}\right).$$

For $a, b \in \mathbb{N}$ with $n \leq \min(a, b)$ and $a \bmod m = b \bmod m$ it holds $a \sim_S b$. Hence, S is recognizable. □

For a monoid $(M, \cdot, e)$ and a subset $S \subseteq M$ the *Kleene star* of S is

$$S^{*(M, \cdot, e)} = \bigcap \{T \mid S \subseteq T \subseteq_{\text{sub}} M\};$$

this is the smallest submonoid of M that contains S. For each set S we have the identity $(\iota_S(S))^{*(S^*, \cdot_S, \lambda_S)} = S^*$. For a monoid $(M, \cdot, e)$ and two subsets $S, T \subseteq M$ the *product* of S and T is

$$S \cdot T = \{s \cdot t \mid s \in S, t \in T\},$$

we also write this as ST. For a commutative monoid, the product of two subsets is often also called the *sum* of these subsets. The product of two languages is also called the *concatenation* of these languages. For a monoid $(M, \cdot, e)$ and a subset $S \subseteq M$ let $S^{0(M, \cdot, e)} = \{e\}$ and for $n \geq 1$ set $S^{n(M, \cdot, e)} = S^{(n-1)(M, \cdot, e)} \cdot S$. This gives us $S^{*(M, \cdot, e)} = \bigcup_{i=0}^{\infty} S^{i(M, \cdot, e)}$. For $k > 0$ and an ordered alphabet $\Sigma = \{a_1, a_2, \ldots, a_k\}$ we call each subset of

$$\{a_1\}^{*(\Sigma^*, \cdot_\Sigma, \lambda_\Sigma)} \{a_2\}^{*(\Sigma^*, \cdot_\Sigma, \lambda_\Sigma)} \cdots \{a_k\}^{*(\Sigma^*, \cdot_\Sigma, \lambda_\Sigma)} \subseteq \Sigma^*$$

a *letter bounded* language.

2.2 Operations

We now consider important operations on subsets of monoids. Each such operation has one or two operands, which are subsets of monoids, and sometimes an additional input from some specified class, for example a monoid homomorphism, and its result is a subset of a monoid. We say that a family of subsets of monoids is closed under an operation if whenever we apply the operation on operands from this family, the result is again in the family. The closure properties of families of subsets of monoids under certain operations have been of great interest since the first days of formal language theory.

Let **FAM** be a family of subsets of monoids, so its elements are pairs of the form $(S, (M, \cdot, e))$ for a monoid $(M, \cdot, e)$ and a subset $S \subseteq M$. We say that **FAM** is

- *closed under union* if for all $(S, (M, \cdot, e)), (T, (M, \cdot, e)) \in$ **FAM** we also have that it holds $(S \cup T, (M, \cdot, e)) \in$ **FAM**.

- *closed under intersection* if for all $(S, (M, \cdot, e)), (T, (M, \cdot, e)) \in$ **FAM** we also have that it holds $(S \cap T, (M, \cdot, e)) \in$ **FAM**.

- *closed under complementation* if for all $(S, (M, \cdot, e)) \in$ **FAM** we also have that it holds $(M \setminus S, (M, \cdot, e)) \in$ **FAM**.

- *closed under direct product* if for all

$$(S_1, (M_1, \cdot_1, e_1)), (S_2, (M_2, \cdot_2, e_2)) \in \textbf{FAM}$$

we also have $(S_1 \times S_2, (M_1 \times M_2, \cdot_1 \times \cdot_2, (e_1, e_2))) \in$ **FAM**.

- *closed under Kleene star* if for all $(S, (M, \cdot, e)) \in \textbf{FAM}$ we also have that it holds $(S^{*(M, \cdot, e)}, (M, \cdot, e)) \in \textbf{FAM}$.

- *closed under product* if for all $(S, (M, \cdot, e)), (T, (M, \cdot, e)) \in \textbf{FAM}$ we also have that it holds $(S \cdot T, (M, \cdot, e)) \in \textbf{FAM}$.

- *closed under homomorphism* if for all

$$(S, (M_1, \cdot_1, e_1)), (X, (M_2, \cdot_2, e_2)) \in \textbf{FAM}$$

and monoid homomorphisms $h : M_1 \to M_2$, we also have that it holds $(h(S), (M_2, \cdot_2, e_2)) \in \textbf{FAM}$.

- *closed under inverse homomorphism* if for all

$$(X, (M_1, \cdot_1, e_1)), (S, (M_2, \cdot_2, e_2)) \in \textbf{FAM}$$

and monoid homomorphisms $h : M_1 \to M_2$, we have

$$\left(h^{-1}(S), (M_1, \cdot_1, e_1) \right) \in \textbf{FAM}.$$

We recall the closure properties of families of recognizable subsets of monoids under the given operations. Let $\textbf{REC}$ be the family of recognizable subsets of monoids.

Proposition 2.7. *The family* $\textbf{REC}$ *is closed under union, intersection, complementation, direct product, and inverse homomorphism.*

Proof. Let $(M, \cdot, e)$ be a monoid and $S, T \subseteq M$ be recognizable subsets. So, there exist finite monoids $(M_0, \cdot_0, e_0)$ and $(M_1, \cdot_1, e_1)$ and monoid homomorphisms $f : M \to M_0$ and $g : M \to M_1$ such that $S = f^{-1}(f(S))$ and $T = g^{-1}(g(T))$. It holds

$$f^{-1}(f(M \setminus S)) = f^{-1}(f(M) \setminus f(S)) = M \setminus S.$$

Thus, $\textbf{REC}$ is closed under complementation. The map $f \times g : M \to M_0 \times M_1$, $a \mapsto (f(a), g(a))$ is a monoid homomorphism. We have

$$(f \times g)^{-1}((f \times g)(S \cup T))$$
$$= (f \times g)^{-1} \left((f \times g)(M) \cap (f(S) \times M_1 \cup M_0 \times g(T)) \right) = S \cup T$$

and

$$(f \times g)^{-1}((f \times g)(S \cap T)) = (f \times g)^{-1}((f \times g)(M) \cap f(S) \times g(T)) = S \cap T.$$

Hence, **REC** is closed under union and intersection. Let $(M_2, \cdot_2, e_2)$ be a monoid and $h : M_2 \to M$ be a monoid homomorphism. We get

$$(f \circ h)^{-1}((f \circ h)(h^{-1}(S))) = h^{-1}(f^{-1}(f(h(M_2) \cap S)))$$
$$= h^{-1}(f^{-1}(f(h(M_2))) \cap S) = (f \circ h)^{-1}((f \circ h)(M_2)) \cap h^{-1}(S) = h^{-1}(S).$$

Therefore, **REC** is closed under inverse homomorphism. Let $(M_3, \cdot_3, e_3)$ be a monoid and $U \subseteq M_3$ be a recognizable subset. The map $F : M \times M_3 \to M_0$, $(a, b) \mapsto f(a)$ is a monoid homomorphism. It holds $F^{-1}(F(S \times M_3)) = F^{-1}(f(S)) = S \times M_3$, which implies that $S \times M_3$ is a recognizable subset of $M \times M_3$. Analogously, $M \times U$ is a recognizable subset of $M \times M_3$. So, $S \times M_3 \cap M \times U = S \times U$ is recognizable, which shows that **REC** is closed under direct product. $\qquad\square$

For a monoid $(M, \cdot, e)$ let $\mathbf{REC}_{(M, \cdot, e)}$ be the family of recognizable subsets of M.

Proposition 2.8. *For a set S the family $\mathbf{REC}_{(\mathrm{Map}_{\mathrm{fin}}(S, \mathbb{N}), +_S, 0_S)}$ of recognizable subsets of the free commutative monoid over S is closed under sum.*

Proof. Let S be a set and T_1 and T_2 be recognizable subsets of $\mathrm{Map}_{\mathrm{fin}}(S, \mathbb{N})$. Since the image of a commutative monoid under a monoid homomorphism is a commutative monoid, for each $i \in \{1, 2\}$ there exist a finite commutative monoid $(M_i, +_i, 0_i)$ and a monoid homomorphism $f_i : \mathrm{Map}_{\mathrm{fin}}(S, \mathbb{N}) \to M_i$ such that $T_i = f_i^{-1}(f_i(T_i))$. Consider the finite commutative monoid

$$\left(2^{M_1 \times M_2}, +_1 \times +_2, \{(0_1, 0_2)\}\right).$$

By the universal property of the free commutative monoid over S, there is a unique monoid homomorphism $f : \mathrm{Map}_{\mathrm{fin}}(S, \mathbb{N}) \to 2^{M_1 \times M_2}$ such that for each $a \in S$ we have

$$(f \circ \kappa_S)(a) = \{((f_1 \circ \kappa_S)(a), 0_2), (0_1, (f_2 \circ \kappa_S)(a))\}.$$

For $g \in \mathrm{Map}_{\mathrm{fin}}(S, \mathbb{N})$ set $|g| = \Sigma_{a \in S}\, g(a)$.

We prove *via* induction over $|g|$ that for all $g \in \mathrm{Map}_{\mathrm{fin}}(S, \mathbb{N})$ it holds

$$f(g) = \{\, (f_1(g_1), f_2(g_2)) \mid g_1, g_2 \in \mathrm{Map}_{\mathrm{fin}}(S, \mathbb{N}),\ g = g_1 +_S g_2 \,\}.$$

This is obvious for $g = 0_S$, so let $g \neq 0_S$. There are $a \in S$ and $h \in \mathrm{Map}_{\mathrm{fin}}(S, \mathbb{N})$ with $g = h +_S \kappa_S(a)$. We get

$$
\begin{aligned}
&f(g) \\
={}& f(h)(+_1 \times +_2) f(\kappa_S(a)) \\
\overset{\text{i.h.}}{=}{}& \{\, (f_1(h_1), f_2(h_2)) \mid h_1, h_2 \in \mathrm{Map}_{\mathrm{fin}}(S, \mathbb{N}),\ h = h_1 +_S h_2 \,\} \\
& (+_1 \times +_2)\{((f_1 \circ \kappa_S)(a), 0_2), (0_1, (f_2 \circ \kappa_S)(a))\} \\
={}& \{\, (f_1(h_1 +_S \kappa_S(a)), f_2(h_2)) \mid h_1, h_2 \in \mathrm{Map}_{\mathrm{fin}}(S, \mathbb{N}),\ h = h_1 +_S h_2 \,\} \\
& \cup \{\, (f_1(h_1), f_2(h_2 +_S \kappa_S(a))) \mid h_1, h_2 \in \mathrm{Map}_{\mathrm{fin}}(S, \mathbb{N}),\ h = h_1 +_S h_2 \,\} \\
={}& \{\, (f_1(g_1), f_2(g_2)) \mid g_1, g_2 \in \mathrm{Map}_{\mathrm{fin}}(S, \mathbb{N}),\ g = g_1 +_S g_2 \,\}.
\end{aligned}
$$

Now, we have

$$
\begin{aligned}
&f^{-1}(f(T_1 + T_2)) \\
={}& f^{-1}\left(\{\, U \in f\left(\mathrm{Map}_{\mathrm{fin}}(S, \mathbb{N})\right) \mid U \cap f_1(T_1) \times f_2(T_2) \neq \emptyset \,\}\right) = T_1 + T_2.
\end{aligned}
$$

Hence, $T_1 + T_2$ is a recognizable subset of $\mathrm{Map}_{\mathrm{fin}}(S, \mathbb{N})$. $\qquad\square$

The following example was given by Shmuel Winograd.

Proposition 2.9. *There is a commutative monoid $(M, +, 0)$ with an element $a \in M$ such that $\{a\}$ is recognizable, but $\{a + a\}$ is not recognizable.*

Proof. Let $(\mathbb{Z}, +, 0)$ be the commutative monoid of integers. Consider the commutative monoid $(M = \mathbb{Z} \cup \{0', a\}, +', 0')$ where $+'$ is defined as follows. For $b, c \in \mathbb{Z}$ let $b +' c = b + c$. For $b \in M$ let $b +' 0' = 0' +' b = b$. For $b \in \mathbb{Z}$ let $b +' a = a +' b = b$. Set $a +' a = 0$. For all $b, c \in \mathbb{Z}$ we have $b \sim_{\{a\}} c$, which gives us $|M/\!\sim_{\{a\}}| \leq 3$. However, in M for all $b, c \in \mathbb{Z}$ with $b \neq c$ it holds $b \not\sim_{\{0\}} c$, which implies $|M/\!\sim_{\{0\}}| = \infty$. $\qquad\square$

Proposition 2.10. *The family* $\mathbf{REC}_{(\mathbb{N}^2,+,\vec{0})}$ *is not closed under Kleene star. The family* $\mathbf{REC}$ *is not closed under homomorphisms mapping from* $\mathbb{N}$ *to* $\mathbb{N}^2$.

Proof. We have $S := \{(1,1)\}^{*(\mathbb{N}^2,+,\vec{0})} = \{\,(n,n) \mid n \in \mathbb{N}\,\}$. Consider the monoid homomorphism $h : \mathbb{N} \to \mathbb{N}^2$, $n \mapsto (n,n)$. It holds $h(\mathbb{N}) = S$. Clearly, $\{(1,1)\}$ in $\mathbb{N}^2$ and $\mathbb{N}$ in $\mathbb{N}$ are recognizable. For all $n, m \in \mathbb{N}$ with $n \neq m$ we have $(n,0) \nsim_S (m,0)$. It follows $|\mathbb{N}^2/\!\!\sim_S| = \infty$. □

For an alphabet Σ the recognizable subsets of Σ^* are called the *regular languages over* Σ. The family of all regular languages is denoted by $\mathbf{REG}$. Let $\mathbf{FAM}$ be a family of languages, so it consists of pairs (L, Σ) of an alphabet Σ and a language $L \subseteq \Sigma^*$. We say that $\mathbf{FAM}$ is

- *closed under intersection with regular languages* if for all $(L, \Sigma) \in \mathbf{FAM}$ and $(R, \Sigma) \in \mathbf{REG}$ we also have $(L \cap R, \Sigma) \in \mathbf{FAM}$.

- *closed under reversal* if for all $(L, \Sigma) \in \mathbf{FAM}$ we also have that it holds $(\{\, w^R \mid w \in L\,\}, \Sigma) \in \mathbf{FAM}$.

- *closed under Kleene plus* if for all pairs $(L, \Sigma) \in \mathbf{FAM}$ we also have that it holds $\left(\bigcup_{i=1}^\infty L^{i(\Sigma^*, \cdot_\Sigma, \lambda_\Sigma)}, \Sigma \right) \in \mathbf{FAM}$.

- *closed under λ-free homomorphism* if for all $(L, \Sigma) \in \mathbf{FAM}$, alphabets Γ, and monoid homomorphisms $h : \Sigma^* \to \Gamma^*$ such that for all $a \in \Sigma$ it holds $h(a) \neq \lambda_\Gamma$, we also have $(h(L), \Gamma) \in \mathbf{FAM}$.

- *closed under substitution* if for all $(L, \Sigma) \in \mathbf{FAM}$, alphabets Γ, and maps $\sigma : \Sigma \to 2^{\Gamma^*}$ such that for all $a \in \Sigma$ it holds $(\sigma(a), \Gamma) \in \mathbf{FAM}$, we also have that the pair of

$$\bigcup \{\, \sigma(a_1)\sigma(a_2) \cdots \sigma(a_n) \mid n \geq 0,\, a_1, a_2, \ldots, a_n \in \Sigma,\, a_1 a_2 \cdots a_n \in L \,\}$$

and Γ is in $\mathbf{FAM}$.

- *closed under permutation closure* if for all $(L, \Sigma) \in \mathbf{FAM}$ we also have that it holds $(\mathsf{perm}(L), \Sigma) \in \mathbf{FAM}$.

We state the missing closure properties of $\mathbf{REG}$, which are well-known, without proof.

Proposition 2.11. *The family of regular languages is closed under reversal, Kleene star, Kleene plus, concatenation, and substitution. However,* **REG** *is not closed under permutation closure.*

2.3 Turing Machines

A *deterministic Turing machine* is a system $M = (Q, \Sigma, \Gamma, \delta, s_0, B, F)$, where Q is the finite set of *internal states*, Γ is a finite set, the *tape alphabet*, $\Sigma \subset \Gamma$ is the non-empty *input alphabet*, $s_0 \in Q$ is the *initial state*, $B \in \Gamma \setminus \Sigma$ is the *blank symbol*, $F \subseteq Q$ is the set of *final states*, and $\delta : Q \times \Gamma \to Q \times \Gamma \times \{-1, 0, 1\}$ is the partial *transition function*. A *configuration* of M is an element from $\Gamma^* \times Q \times (\Gamma^* \setminus \{\lambda_\Gamma\})$. A Turing machine works on an tape, which is divided into infinitely many cells. In each cell a symbol from Γ is stored. The current situation of the tape can be modelled by a map $\mathbb{Z} \to \Gamma$ such that for only a finite number of arguments the value is not B. The machine has a head which can read and overwrite the symbols in the cells. When the machine M is in configuration (v, s, w), this means that the current internal state of the machine is s, the part of the tape which was already visited contains the string vw, and the head is on the first symbol of w. Each cell of the tape which was not visited yet contains the symbol B. At the beginning of the computation, the machine is in the configuration (λ_Γ, s_0, w), where w is the input word. If the input word is empty, we have $w = B$. In each step of the computation the head reads the symbol on which it currently stands and the transition function tells us to which internal state the machine switches, by which tape symbol the symbol which was read in this step is overwritten, and if the head moves one cell to the left, one cell to the right, or does not move. The machine halts, when δ is undefined for the current configuration. If we define a one-to-one correspondence between $\mathbb{N}$ and Σ^*, M computes a partial function $f : \mathbb{N} \to \mathbb{N}$, where $f(n)$ is undefined if M does not halt on input n or the output is invalid. The machine M also works as an acceptor of a language $L(M) \subseteq \Sigma^*$. A word w is accepted by M if M halts in a final state on input w.

For a *nondeterministic Turing machine* the transition function maps to $2^{Q \times \Gamma \times \{-1,0,1\}}$. In each step of the computation the machine chooses one of the possible successor configurations. A word w is accepted by the nondeterministic machine if there is a computation on input w that halts in a final state. It can be shown that nondeterministic Turing machines can only accept the same languages as deterministic ones. There are also definitions of Turing machines with separate input, output, and working tape, with multiple working tapes, or with a working tape that is bounded on one side. For each variant each Turing machine has exactly one head for each tape. All these variants of Turing machines have the same computational power. The Church-Turing Thesis says that a Turing machine can compute everything that is computable by any kind of machine. The languages accepted by Turing-machines are called the *recursively enumerable* languages. Chomsky's grammars of type 0 generate exactly the recursively enumerable languages. Because the set of all Turing machines up to isomorphism is countable, there are languages which are not recursively enumerable, even over a unary alphabet.

A *decision problem* is a problem where for each input exactly one of the answers 'yes' or 'no' is correct. So, a decision problem can be seen as a language over a suitable alphabet. A decision problem is called *decidable* if there is a deterministic Turing machine that for every input halts and gives the correct answer. A decision problem which is not decidable is called *undecidable*. By Rice's theorem for a Turing machine M every non-trivial property of $L(M)$ is undecidable. Given an automaton model such that each automaton A accepts a language $L(A)$, the following decision problems are of high interest in formal language theory. Let Σ be an alphabet and consider only automata of the given type with input alphabet Σ.

- The *word problem*: given $w \in \Sigma^*$ and an automaton A, is $w \in L(A)$?

- *Emptiness*: given an automaton A, is $L(A) = \emptyset$?

- *Finiteness*: given an automaton A, is $L(A)$ finite?

- *Universality*: given an automaton A, is $L(A) = \Sigma^*$?

- *Regularity*: given an automaton A, is $L(A)$ regular?

- *Disjointness*: given automata A and B, is $L(A) \cap L(B) = \emptyset$?
- *Inclusion*: given automata A and B, is $L(A) \subseteq L(B)$?
- *Equivalence*: given automata A and B, is $L(A) = L(B)$?

For Turing machines all these problems are undecidable.

A *linear bounded automaton* is a Turing machine for which the tape consists only of the cells on which the input is given. The languages accepted by nondeterministic linear bounded automata are the *context-sensitive* languages. This language family is denoted by **CS**. It consists of all languages generated by Chomsky's grammars of type 1. The problem if nondeterministic linear bounded automata can accept more languages than deterministic ones is still open. There are recursively enumerable languages which are not context-sensitive. For nondeterministic linear bounded automata the word problem is decidable, while emptiness, finiteness, universality, regularity, disjointness, inclusion, and equivalence are undecidable. Closure properties of **CS** are given in Table 2.1.

Example 2.12. The language $\left\{ a^{2^n} \mid n \geq 0 \right\} \subset \{a\}^*$ is accepted by a deterministic linear bounded automaton as follows. The automaton scans the input from left to right and marks every second a. When the head is at the end of the input, it moves back to the beginning, marking every second unmarked a meanwhile. The automaton continues this way marking every second unmarked a in each sweep. If the number of unmarked a's was odd and bigger than one before a sweep, the input is rejected. If there was just one unmarked a left before a sweep, the input is accepted.

The most important complexity classes are usually defined using multi-tape Turing machines in the following way. Let $f : \mathbb{N} \to \mathbb{R}_{\geq 0}$. The class **DTIME**$(f)$ consists of all decision problems for which there is a deterministic multi-tape Turing machine which for every input w halts after $O(f(|w|))$ steps and gives the correct answer. The class **NTIME**(f) consists of all decision problems for which there is a nondeterministic multi-tape Turing machine that accepts exactly those inputs for which the answer is 'yes' such that for every input w every computation halts after $O(f(|w|))$ steps. The class **coNTIME**(f) consists of the complements of the decision problems in **NTIME**(f). The class **DSPACE**(f) consists of all decision problems for which there is a deterministic multi-tape Turing machine which has a read-only input tape such that for every input w the heads visit $O(f(|w|))$ working tape cells during the computation and the machine halts and gives the correct answer. The class **NSPACE**(f) consists of all decision problems for which there is a nondeterministic multi-tape Turing machine which has a read-only input tape and accepts exactly those inputs for which the answer is 'yes' such that for every input w and every computation on that input the heads visit $O(f(|w|))$ working tape cells during the computation and the machine halts. The class **coNSPACE**(f) consists of the complements of the decision problems in **NSPACE**(f).

The binary logarithm function is denoted as log by us. Set

$$\mathsf{L} = \mathbf{DSPACE}(\log(n)), \qquad \mathsf{NL} = \mathbf{NSPACE}(\log(n)),$$
$$\mathsf{P} = \bigcup_{k=1}^{\infty} \mathbf{DTIME}\left(n^k\right), \qquad \mathsf{NP} = \bigcup_{k=1}^{\infty} \mathbf{NTIME}\left(n^k\right),$$
$$\mathsf{coNP} = \bigcup_{k=1}^{\infty} \mathbf{coNTIME}\left(n^k\right), \quad \mathsf{PSPACE} = \bigcup_{k=1}^{\infty} \mathbf{DSPACE}\left(n^k\right),$$
$$\mathsf{EXP} = \bigcup_{k=1}^{\infty} \mathbf{DTIME}\left(2^{n^k}\right), \qquad \mathsf{NEXP} = \bigcup_{k=1}^{\infty} \mathbf{NTIME}\left(2^{n^k}\right),$$
$$\mathsf{coNEXP} = \bigcup_{k=1}^{\infty} \mathbf{coNTIME}\left(2^{n^k}\right).$$

It was shown by Walter J. Savitch in 1970 that

$$\mathsf{PSPACE} = \bigcup_{k=1}^{\infty} \mathbf{NSPACE}\left(n^k\right) \text{ [68]}.$$

Neil Immerman [42] and Róbert Szelepcsényi [72] in 1987 independently proved that $\mathsf{NL} = \mathbf{coNSPACE}(\log(n))$, which implies that **CS** is closed under the operation of complementation. The problems if **NP** = **coNP** and if **NEXP** = **coNEXP** are both still open. We have

$$\mathsf{L} \subseteq \mathsf{NL} \subseteq \mathsf{P} \subseteq \mathsf{NP} \subseteq \mathsf{PSPACE} \subseteq \mathsf{EXP} \subseteq \mathsf{NEXP}.$$

For none of the '$\subseteq$' in the line above it is known whether it can be replaced by '$\subset$'. By the space hierarchy theorems it holds $\mathsf{NL} \subset \mathsf{PSPACE}$. The time hierarchy theorems give us $\mathsf{P} \subset \mathsf{EXP}$ and $\mathsf{NP} \subset \mathsf{NEXP}$. It is still open if $\mathsf{L} \subset \mathsf{NP}$, if $\mathsf{P} \subset \mathsf{PSPACE}$, if $\mathsf{NP} \subset \mathsf{EXP}$, and if $\mathsf{PSPACE} \subset \mathsf{NEXP}$.

A decision problem D is *log-space many-one reducible* to a decision problem E, written as $D \leq_m^{\log} E$, if there is a deterministic Turing machine with a read-only input tape, a working tape, and a one-way, write-only output tape such that for every input w the working tape head visits $O(\log(|w|))$ tape cells during the computation and the machine halts and fulfils that the output is in E if and only if $w \in D$. For decision problems D, E, and F one can show the following. If $D \leq_m^{\log} E$ and $E \leq_m^{\log} F$, then $D \leq_m^{\log} F$. If $D \leq_m^{\log} E$ and $E \in \mathsf{L}$, then $D \in \mathsf{L}$. If $D \leq_m^{\log} E$ and $E \in \mathsf{NL}$, then $D \in \mathsf{NL}$. A decision problem D is called NL-*hard* if for all $E \in \mathsf{NL}$ it holds $E \leq_m^{\log} D$. A decision problem is called NL-*complete* if it is contained in NL and NL-hard. For example, emptiness, finiteness, and universality for deterministic finite automata are NL-complete.

Let $f, g : \mathbb{N} \to \mathbb{R}_{\geq 0}$. We define the class $\mathbf{DST}(f, g)$ to consist of all decision problems for which there is a deterministic multi-tape Turing machine which has a read-only input tape such that for every input w the heads visit $O(f(|w|))$ working tape cells during the computation and the machine halts after $O(g(|w|))$ steps and gives the correct answer. We define the class $\mathbf{NST}(f, g)$ to consist of all decision problems for which there is a nondeterministic multi-tape Turing machine which has a read-only input tape and accepts exactly those inputs for which the answer is 'yes' such that for every input w and every computation on that input the heads visit $O(f(|w|))$ working tape cells during the computation and the machine halts after $O(g(|w|))$ steps.

2.4 Pushdown Automata

A *nondeterministic pushdown automaton* (PDA) is a system

$$A = (Q, \Sigma, \Gamma, \delta, s_0, \bot, F),$$

where Q is the finite set of *internal states*, Σ is the finite non-empty *input alphabet*, Γ is the finite *pushdown alphabet*, $s_0 \in Q$ is the *initial state*, the symbol $\perp \in \Gamma$ is the *initial pushdown symbol*, $F \subseteq Q$ is the set of *final states*, and the *transition function* δ maps from $Q \times (\Sigma \cup \{\lambda_\Sigma\}) \times \Gamma$ to the set of finite subsets of $Q \times \Gamma^*$. A *configuration* of A is an element from $Q \times \Sigma^* \times \Gamma^*$. If the automaton is in configuration (s, w, α), the current internal state of A is s, the unread part of the input is w, and the word α is stored in the pushdown store, where the first symbol of α is at the top of the store. Define the binary relation $\vdash_A$, or just $\vdash$ if it is clear which PDA we are referring to, on the set of configurations of A as follows. For $s, t \in Q$, $a \in \Sigma \cup \{\lambda_\Sigma\}$, $w \in \Sigma^*$, $B \in \Gamma$, and $\alpha, \beta \in \Gamma^*$ with $(t, \alpha) \in \delta(s, a, B)$, we have $(s, aw, B\beta) \vdash (t, w, \alpha\beta)$. The language accepted by A is

$$L(A) = \{ w \in \Sigma^* \mid \exists f \in F, \ \alpha \in \Gamma^* : (s_0, w, \perp) \vdash_A^* (f, \lambda_\Sigma, \alpha) \}.$$

The automaton A is called *deterministic*, a DPDA, if for all $s \in Q$, $a \in \Sigma$, and $B \in \Gamma$ we have $|\delta(s, a, B) \cup \delta(s, \lambda_\Sigma, B)| \leq 1$. If A is a DPDA, we sometimes write δ as a partial function from $Q \times (\Sigma \cup \{\lambda_\Sigma\}) \times \Gamma$ to $Q \times \Gamma^*$. The languages accepted by PDAs are the *context-free* languages. This language family is denoted by **CF** and consists of all languages that are generated by Chomsky's grammars of type 2. The languages accepted by DPDAs are called the *deterministic context-free* languages. That family is denoted by **DCF**. It can be shown that for each PDA a deterministic linear bounded automaton accepting the same language is effectively constructable. This means that the construction can be computed by a Turing machine by using a suitable encoding for the automata.

The following pumping lemma is often used to show that a certain language is not context free. It was given by Yehoshua Bar-Hillel *et al.* in 1961.

Lemma 2.13 ([1]). *Let Σ be an alphabet and $L \subseteq \Sigma^*$ be context free. Then, there is an $n \geq 1$ such that for all $z \in L$ with $|z| \geq n$ there are $u, v, w, x, y \in \Sigma^*$ with $z = uvwxy$, $|vx| \geq 1$, and $|vwx| \leq n$ such that for all $i \geq 0$ it holds $uv^i wx^i y \in L$.*

Example 2.14. It follows from Lemma 2.13 that the language $\{ a^{2^n} \mid n \geq 0 \} \subset \{a\}^*$ from Example 2.12 is not context free.

For deterministic context-free languages a pumping lemma was given by Sheng Yu in 1989:

Lemma 2.15 ([76]). *Let Σ be an alphabet and $L \subseteq \Sigma^*$ be deterministic context free. Then, there is an $n \geq 1$ such that for all $\sigma \in \Sigma$ and $x, y, z \in \Sigma^*$ with $|x| > n$ and $x\sigma y, x\sigma z \in L$ at least one of the following two conditions is true.*

- *There are words $x_1, x_2, x_3, x_4, x_5 \in \Sigma^*$ fulfilling $x = x_1 x_2 x_3 x_4 x_5$, $|x_2 x_4| \geq 1$, and $|x_2 x_3 x_4| \leq n$ such that for all numbers $i \geq 0$ it holds $x_1 x_2^i x_3 x_4^i x_5 \sigma y \in L$ and $x_1 x_2^i x_3 x_4^i x_5 \sigma z \in L$.*

- *There are $x_1, x_2, x_3, y_1, y_2, y_3, z_1, z_2, z_3 \in \Sigma^*$ with $x = x_1 x_2 x_3$, $\sigma y = y_1 y_2 y_3$, $\sigma z = z_1 z_2 z_3$, $|x_2| \geq 1$, and $|x_2 x_3| \leq n$ such that for all $i \geq 0$ it holds $x_1 x_2^i x_3 y_1 y_2^i y_3 \in L$ and $x_1 x_2^i x_3 z_1 z_2^i z_3 \in L$.*

Example 2.16. Consider the DPDA

$$A = (\{s_0, s_a, s_b\}, \{a, b\}, \{\bot, C\}, \delta, s_0, \bot, \{s_0\}),$$

where the map δ is defined as follows. For all $c, d \in \{a, b\}$ with $c \neq d$ we have

$$\delta(s_0, c, \bot) = \{(s_c, C\bot)\}, \qquad \delta(s_c, c, C) = \{(s_c, CC)\},$$
$$\delta(s_c, d, C) = \{(s_c, \lambda)\}, \qquad \delta(s_c, \lambda, \bot) = \{(s_0, \bot)\}.$$

For all arguments which are not covered by the last sentence the value is the empty set. It is not hard to see that $L(A) = \{\, w \in \{a, b\}^* \mid |w|_a = |w|_b \,\}$. For $n, m \geq 0$ with $n \neq m$ we have $a^n \not\approx_{L(A)} a^m$. Hence, $L(A)$ is not regular. Now, we look at the DPDA

$$B = (\{s_0, s_a, s_a', s_b\}, \{a, b\}, \{\bot, C\}, \gamma, s_0, \bot, \{s_0\}),$$

where γ is given in the following. We have

$$\gamma(s_0, b, \bot) = \{(s_b, CC\bot)\}, \qquad \gamma(s_b, b, C) = \{(s_b, CCC)\},$$
$$\gamma(s_a, b, C) = \{(s_a', \lambda)\}, \qquad \gamma(s_a', \lambda, C) = \{(s_a, \lambda)\},$$
$$\gamma(s_a', \lambda, \bot) = \{(s_b, C\bot)\}.$$

For all other arguments from $\{s_0, s_a, s_b\} \times \{a, b, \lambda\} \times \{\bot, C\}$, the map γ agrees with δ. For all arguments which are not covered yet the value is the empty set. This gives us $L(B) = \{\, w \in \{a,b\}^* \mid |w|_a = 2 \cdot |w|_b \,\}$. It is easy to see that **CF** is closed under union, so $L(A) \cup L(B)$ is context free. To see that $L(A) \cup L(B)$ is not deterministic context free, we set $\sigma = a$, $x = b^{n+1}a^n$, $y = \lambda$, and $z = a^{n+1}$ in Lemma 2.15.

The word problem, emptiness, and finiteness are decidable for PDAs. Disjointness and inclusion are undecidable even for DPDAs. Universality, regularity, and equivalence are undecidable for PDAs, but decidable for DPDAs. Decidability of equivalence for DPDAs was proven by Géraud Sénizergues in 1997 [70]. Closure properties of **CF** and **DCF** are given in Table 2.1.

A PDA $(Q, \Sigma, \Gamma, \delta, s_0, \perp, F)$ is called λ-*free* if for all $(q, A) \in Q \times \Gamma$ it holds $\delta(q, \lambda_\Sigma, A) = \emptyset$. For each PDA there is an equivalent λ-free PDA. However, there exists a DPDA for which there is no equivalent λ-free DPDA, see the book *Introduction to Formal Language Theory* by Harrison [37]. The following pumping lemma for λ-free deterministic pushdown automata was given by Yoshihide Igarashi. It makes use of the Myhill-Nerode relation, which is defined as follows. For an alphabet Σ and a language $L \subseteq \Sigma^*$ the *Myhill-Nerode equivalence relation* $\simeq_L$ on Σ^* is defined as follows. For $v, w \in \Sigma^*$ we have $v \simeq_L w$ if for all $x \in \Sigma^*$ the condition $vx \in L$ is equivalent to the condition $wx \in L$.

Lemma 2.17 ([41]). *Let Σ be an alphabet, $L \subseteq \Sigma^*$, and $f : \mathbb{N} \to \mathbb{N}$ be a function such that for all $n, m \in \mathbb{N}_{>0}$ there are $x_1, x_2, \ldots, x_n \in \Sigma^*$,*

$$y_{1,1}, y_{1,2}, \ldots, y_{1,m}, y_{2,1}, y_{2,2}, \ldots, y_{2,m}, \ldots \ldots, y_{n,1}, y_{n,2}, \ldots, y_{n,m} \in \Sigma^*,$$

and

$$w_{1,1}, w_{1,2}, \ldots, w_{1,m}, w_{2,1}, w_{2,2}, \ldots, w_{2,m}, \ldots \ldots, w_{n,1}, w_{n,2}, \ldots, w_{n,m} \in \Sigma^*$$

such that

- *for all numbers $i \in \{1, 2, \ldots, n\}$ and $j_1, j_2 \in \{1, 2, \ldots, m\}$ fulfilling $j_1 < j_2$ it holds $x_i y_{i,j_1} \not\simeq_L x_i y_{i,j_2}$,*
- *for all $i_1, i_2 \in \{1, 2, \ldots, n\}$ with $i_1 < i_2$, $j_1, j_2 \in \{1, 2, \ldots, m\}$, and prefixes z_1 of y_{i_1,j_1} and z_2 of y_{i_2,j_2} we have $x_{i_1} z_1 \not\simeq_L x_{i_2} z_2$,*

- *for all $i \in \{1, 2, \ldots, n\}$ and $j \in \{1, 2, \ldots, m\}$ it holds $|w_{i,j}| \leq f(n)$ and $x_i y_{i,j} w_{i,j} \in L$, and*

- *for all $i \in \{1, 2, \ldots, n\}$, $j \in \{1, 2, \ldots, m\}$, and $u_1, u_2, u_3 \in \Sigma^*$ with $x_i = u_1 u_2 u_3$ and $|u_2| > 0$ there exists a number $t \geq 0$ such that $u_1 u_2^t u_3 y_{i,j} w_{i,j} \notin L$.*

Then, L is not accepted by a λ-free deterministic pushdown automaton.

A PDA $(Q, \Sigma, \Gamma = \{\bot, A\}, \delta, s_0, \bot, F)$ is called a *counter automaton* if for all $q \in Q$ and $a \in \Sigma \cup \{\lambda_\Sigma\}$ it holds $\delta(q, a, A) \subset Q \times \{A\}^{*(\Gamma^*, \Gamma, \lambda_\Gamma)}$ and

$$\delta(q, a, \bot) \subset Q \times \left(\{\lambda_\Gamma\} \cup \{A\}^{*(\Gamma^*, \Gamma, \lambda_\Gamma)} \cdot_\Gamma \{\bot\} \right).$$

It is not hard to see that for each counter automaton there is an equivalent λ-free counter automaton.

A PDA $(Q, \Sigma, \Gamma, \delta, s_0, \bot, F)$ is called a *one-turn pushdown automaton* if there are no states $p, q, r \in Q$ and words $u, v, w \in \Sigma^*$ and $x, y, z \in \Gamma^*$ such that

$$(s_0, u, \bot) \vdash^* (p, v, x) \vdash^* (q, w, y) \vdash^* (r, \lambda, z)$$

and $|x| > |y| < |z|$. The languages accepted by one-turn pushdown automaton are the languages generated by linear context-free grammars. These languages are called the *linear* languages. It was shown by Andrzej Ehrenfeucht *et al.* in 1982 that each permutation closed linear language is regular [23].

2.5 Finite Automata

We define a *nondeterministic finite automaton with multiple start states and spontaneous or λ-transitions*, a λ-MNFA for short, as a tuple $A = (Q, \Sigma, \delta, S, F)$, where Q is the finite set of *internal states*, Σ is the finite non-empty *input alphabet*, $\delta : Q \times (\Sigma \cup \{\lambda_\Sigma\}) \to 2^Q$ is the *transition function*, $S \subseteq Q$ is the set of *start states*, and $F \subseteq Q$ is the set of *final states*. For $p \in Q$, $a \in \Sigma \cup \{\lambda_\Sigma\}$, and $q \in \delta(p, a)$ we sometimes write that the rule $pa \to q$ is in δ. A *configuration* of A is an element from $Q \times \Sigma^*$. Define the binary relation $\vdash_A$, or just $\vdash$ if it is clear which λ-MNFA we are referring to, on the set of configurations of A as follows. For $p \in Q$, $a \in \Sigma \cup \{\lambda_\Sigma\}$, $q \in \delta(p, a)$, and $w \in \Sigma^*$ we have $(p, aw) \vdash (q, w)$. The language accepted by A is

$$L(A) = \{\, w \in \Sigma^* \mid \exists s \in S,\, f \in F : (s, w) \vdash_A^* (f, \lambda_\Sigma) \,\}.$$

For all $q \in Q$ we set $\Sigma_{q,\delta} = \{\, a \in \Sigma \mid \delta(q, a) \neq \emptyset \,\}$. If it is clear which λ-MNFA we are referring to, we also just write Σ_q for $\Sigma_{q,\delta}$.

The automaton A is called a *nondeterministic finite automaton with multiple start states*, an MNFA, if for all $p \in Q$ the set $\delta(p, \lambda_\Sigma)$ is empty. We call A a *nondeterministic finite automaton with λ-transitions*, a λ-NFA, if $|S| = 1$. If A is an MNFA and a λ-NFA, we say that A is a *nondeterministic finite automaton*, an NFA. The automaton A is said to be a *deterministic finite automaton with multiple start states*, an MDFA, if A is an MNFA and for all $p \in Q$ and $a \in \Sigma$ we have $|\delta(p, a)| \leq 1$. If A is an MDFA, we sometimes write δ as a partial function $Q \times \Sigma \to Q$. We call A a *deterministic finite automaton*, a DFA, if it is an MDFA and an NFA. Clearly, each language accepted by a DFA is in **DCF**. If A is a DFA and the transition function $\delta : Q \times \Sigma \to Q$ is a total function, we call A a *deterministic finite automaton with total transition function*, a DFAwttf. By the powerset construction all of the models mentioned in this section have the same computational power:

Theorem 2.18. *Let A be a λ-MNFA. Then, one can effectively construct a DFAwttf B such that $L(B) = L(A)$.*

Proof. Let $A = (Q, \Sigma, \delta, S, F)$ and define the DFA B as

$$\left(2^Q, \Sigma, \Delta, \{\{\, p \in Q \mid \exists s \in S : (s, \lambda_\Sigma) \vdash_A^* (p, \lambda_\Sigma) \,\}\}, \right.$$

$$\left. \{\, P \subseteq Q \mid P \cap F \neq \emptyset \,\} \right),$$

where the total transition function $\Delta : 2^Q \times \Sigma \to 2^Q$ is defined by

$$(P, a) \mapsto \{\, q \in Q \mid \exists p \in P : (p, a) \vdash_A^* (q, \lambda_\Sigma) \,\}.$$

By induction over $|w|$, one easily sees that for all $P, R \subseteq Q$ and $w \in \Sigma^* \setminus \{\lambda_\Sigma\}$ with $(P, w) \vdash_B^* (R, \lambda_\Sigma)$ it holds

$$R = \{\, q \in Q \mid \exists p \in P : (p, w) \vdash_A^* (q, \lambda_\Sigma) \,\}.$$

The language $L(B)$ consists of all $w \in \Sigma^*$ which fulfil

$$\exists f \in F, \, P \subseteq Q :$$
$$(f \in P \wedge (\{\, p \in Q \mid \exists s \in S : (s, \lambda_\Sigma) \vdash_A^* (p, \lambda_\Sigma) \,\}, w) \vdash_B^* (P, \lambda_\Sigma)).$$

That is equivalent to the condition

$$\exists f \in F : \exists p \in Q : ((p, w) \vdash_A^* (f, \lambda_\Sigma) \wedge \exists s \in S : (s, \lambda_\Sigma) \vdash_A^* (p, \lambda_\Sigma)).$$

This gives us

$$L(B) = \{\, w \in \Sigma^* \mid \exists f \in F, \, s \in S : (s, w) \vdash_A^* (f, \lambda_\Sigma) \,\} = L(A).$$

$\square$

We call a DFAwttf A *minimal* if for all DFAswttf B over the same input alphabet with $L(B) = L(A)$ the number of internal states of B is not smaller than the number of internal states of A. For two DFAswttf $A = (Q, \Sigma, \delta, \{s_0\}, F)$ and $A' = (Q', \Sigma, \delta', \{s_0'\}, F')$ an *isomorphism* between A and A' is a bijection $f : Q \to Q'$ such that $f(s_0) = s_0'$, $f(F) = F'$, and for all $q \in Q$ and $a \in \Sigma$ it holds $f(\delta(q, a)) = \delta'(f(q), a)$.

Theorem 2.19. *Let Σ be an alphabet and $L \subseteq \Sigma^*$. Then, L is accepted by a DFAwttf if and only if $\Sigma^*/\simeq_L$ is finite. If $\Sigma^*/\simeq_L$ is finite, the number of internal states of any minimal DFAwttf over the input alphabet Σ accepting L equals $|\Sigma^*/\simeq_L|$. Any two minimal DFAswttf over the same input alphabet accepting the same language are isomorphic. Given a DFAwttf, one can effectively construct a minimal DFAwttf accepting the same language.*

Proof. Assume that there is a DFAwttf $A = (Q, \Sigma, \delta, \{s_0\}, F)$ fulfilling $L(A) = L$. For each $w \in \Sigma^*$ there is a unique state $q_w \in Q$ such that $(s_0, w) \vdash_A^* (q_w, \lambda_\Sigma)$. Let $v, w \in \Sigma^*$ with $q_v = q_w$. For all $x \in \Sigma^*$

it clearly holds $q_{vx} = q_{wx}$. So, we get $v \simeq_L w$. That implies $|\Sigma^*/\simeq_L| \leq |Q| < \infty$.

Assume now that $\Sigma^*/\simeq_L$ is finite. Consider the DFA

$$B = (\Sigma^*/\simeq_L, \Sigma, \gamma, \{[\lambda_\Sigma]_{\simeq_L}\}, \{[w]_{\simeq_L} \mid w \in L\}),$$

where the total transition function $\gamma : (\Sigma^*/\simeq_L) \times \Sigma \to \Sigma^*/\simeq_L$ is given by $([w]_{\simeq_L}, a) \mapsto [wa]_{\simeq_L}$. This map is well-defined because for all $v, w \in \Sigma^*$ with $v \simeq_L w$ and $a \in \Sigma$ it holds $va \simeq_L wa$. For each $w \in \Sigma^*$ we have $([\lambda_\Sigma]_{\simeq_L}, w) \vdash_B^* ([w]_{\simeq_L}, \lambda_\Sigma)$. It follows $L(B) = L$. What we have shown above implies that B is minimal. Let now $A = (Q, \Sigma, \delta, \{s_0\}, F)$ be a minimal DFAwttf such that $L(A) = L$. For each $w \in \Sigma^*$ let $q_w \in Q$ be defined as above. Because of $|Q| = |\Sigma^*/\simeq_L|$, for all $v, w \in \Sigma^*$ the condition $v \simeq_L w$ is equivalent to the condition $q_v = q_w$. The map $\Sigma^*/\simeq_L \to Q$, $[w]_{\simeq_L} \mapsto q_w$ is an isomorphism between B and A.

Let $C = (P, \Sigma, \beta, \{t_0\}, G)$ be a DFAwttf. For each $w \in \Sigma^*$ there is a unique state $p_w \in P$ such that $(t_0, w) \vdash_C^* (p_w, \lambda_\Sigma)$. Assume that there are $v, w \in \Sigma^*$ such that $v \simeq_{L(C)} w$ does not hold. W.l.o.g. there is a $u \in \Sigma^*$ with $p_{vu} \in G$ and $p_{wu} \notin G$. If $|u| \geq |P|^2$, there are $x, y, z \in \Sigma^*$ with $u = xyz$, $y \neq \lambda_\Sigma$, and $(p_{vx}, p_{wx}) = (p_{vxy}, p_{wxy})$. It follows $p_{vxz} = p_{vu}$ and $p_{wxz} = p_{wu}$. Iterating this, we get a $u' \in \Sigma^*$ with $|u'| < |P|^2$, $p_{vu'} = p_{vu}$, and $p_{wu'} = p_{wu}$. Thus, we have shown that for all $v, w \in \Sigma^*$ the condition $v \simeq_{L(C)} w$ is equivalent to the condition that for all $u \in \Sigma^*$ with $|u| < |P|^2$ it holds $(p_{vu}, p_{wu}) \in G \times G \cup (P \backslash G) \times (P \backslash G)$. This implies that a minimal DFAwttf accepting $L(C)$ is effectively constructable. $\quad\square$

It is not hard to show that for DFAswttf the word problem, emptiness, finiteness, universality, regularity, disjointness, inclusion, and equivalence are decidable. By Theorem 2.18 these problems are also decidable for λ-MNFAs. From Theorem 2.19 we can deduce:

Corollary 2.20. *The languages accepted by DFAswttf are the regular languages.*

Proof. Let Σ be an alphabet and $L \subseteq \Sigma^*$. For all $v, w \in \Sigma^*$ we have that $v \sim_L w$ implies $v \simeq_L w$. If L is a regular language, it holds $|\Sigma^*/\simeq_L| \leq$

$|\Sigma^*/\sim_L| < \infty$, which, together with Theorem 2.19, gives us that L is accepted by a DFAwttf. Assume now that L is accepted by a DFAwttf. By Theorem 2.19 it holds $|\Sigma^*/\simeq_L| < \infty$. Now, we consider the finite monoid $\left((\Sigma^*/\simeq_L)^{\Sigma^*/\simeq_L}, \circ_R, \mathrm{id}_{\Sigma^*/\simeq_L}\right)$, where $\circ_R$ is defined as follows. For functions $f, g \in (\Sigma^*/\simeq_L)^{\Sigma^*/\simeq_L}$ and $\alpha \in \Sigma^*/\simeq_L$ we have $(f \circ_R g)(\alpha) = g(f(\alpha))$. The map $h : \Sigma^* \to (\Sigma^*/\simeq_L)^{\Sigma^*/\simeq_L}$ given by $w \mapsto ([v]_{\simeq_L} \mapsto [vw]_{\simeq_L})$ is a monoid homomorphism such that

$$h^{-1}(h(L)) = h^{-1}\left(h(\Sigma^*) \cap \left\{ f \in (\Sigma^*/\simeq_L)^{\Sigma^*/\simeq_L} \;\middle|\; f([\lambda_\Sigma]_{\simeq_L}) \in L/\simeq_L \right\}\right),$$

which equals L. Hence, L is regular. $\qquad\square$

Example 2.21. Consider the DFAwttf $A = (\{p, q, r\}, \{a, b\}, \delta, \{p\}, \{p\})$ with $\delta(p, a) = q$, $\delta(q, b) = p$, and the value of δ is r for all other arguments. It is not hard to see that $L(A) = \{ (ab)^n \mid n \geq 0 \} \subset \{a, b\}^*$. In Examples 2.1 and 2.4 we have already seen that for the syntactic monoid of $L(A)$ it holds $\left|\{a, b\}^*/\sim_{L(A)}\right| = 6$. For the equivalence classes of the Myhill-Nerode relation $\simeq_{L(A)}$ we get that

$$\{a, b\}^*/\simeq_{L(A)} = \left\{ [\lambda_{\{a,b\}}]_{\simeq_{L(A)}}, [a]_{\simeq_{L(A)}}, [b]_{\simeq_{L(A)}} \right\}$$

and that this three classes are pairwise different. Thus, A is minimal.

Two other famous formalisms for describing regular languages are Chomsky's grammars of type 3 and Kleene's regular expressions.

2.6 Semilinear Sets

Let $k \geq 0$. A subset of $\mathbb{N}^k$ is called *linear* if it is of the form

$$\mathsf{L}(\vec{c}, P) = \left\{ \vec{c} + \sum_{\vec{p} \in P} \lambda_{\vec{p}} \cdot \vec{p} \;\middle|\; \forall \vec{p} \in P : \lambda_{\vec{p}} \in \mathbb{N} \right\}$$

for a finite set $P \subseteq \mathbb{N}^k$ and a $\vec{c} \in \mathbb{N}^k$, where P is called the set of *periods* and $\vec{c}$ is called the *constant vector* of this representation. A subset of $\mathbb{N}^k$ is called *semilinear* if it is a finite union of linear subsets of $\mathbb{N}^k$. Let $k \geq 0$. A *semilinear representation in* $\mathbb{N}^k$ is a triple $(I, (\vec{c}_i)_{i \in I}, (P_i)_{i \in I})$, where I is a finite set and for each $i \in I$ we have $\vec{c}_i \in \mathbb{N}^k$ and that $P_i \subseteq \mathbb{N}^k$ is finite. The semilinear subset of $\mathbb{N}^k$ represented by a semilinear representation in $\mathbb{N}^k$, given as $R = (I, (\vec{c}_i)_{i \in I}, (P_i)_{i \in I})$, is $S(R) = \cup_{i \in I} \mathsf{L}(\vec{c}_i, P_i)$. We use a notion from Ginsburg and Spanier [32]: for finite $C, P \subseteq \mathbb{N}^k$ let

$$\mathsf{L}(C, P) = \left\{ \vec{c} + \sum_{\vec{p} \in P} \lambda_{\vec{p}} \cdot \vec{p} \;\middle|\; \vec{c} \in C, \, \forall \vec{p} \in P : \lambda_{\vec{p}} \in \mathbb{N} \right\},$$

where we call C the set of *constants* and P the set of *periods*. Then, for finite $C, P \subseteq \mathbb{N}^k$ we have $\mathsf{L}(C, P) = \cup_{\vec{c} \in C} \mathsf{L}(\vec{c}, P)$. Following Dmitry Chistikov and Christoph Haase, we call sets of the form $\mathsf{L}(C, P)$ for finite $C, P \subseteq \mathbb{N}^k$ *hybrid linear subsets of* $\mathbb{N}^k$ [15]. A *representation through hybrid linear sets in* $\mathbb{N}^k$ is a triple $(I, (C_i)_{i \in I}, (P_i)_{i \in I})$, where I is a finite set and for each $i \in I$ we have $C_i, P_i \subseteq \mathbb{N}^k$ with $|C_i \cup P_i| < \infty$. The semilinear subset of $\mathbb{N}^k$ represented by a representation through hybrid linear sets in $\mathbb{N}^k$, given as $R = (I, (C_i)_{i \in I}, (P_i)_{i \in I})$, is $S(R) = \cup_{i \in I} \mathsf{L}(C_i, P_i)$. Clearly, each semilinear representation can be seen as a special case of a representation through hybrid linear sets. On the other hand, each representation through hybrid linear sets can easily be transformed to a semilinear representation of the same set.

Ginsburg and Spanier proved a Carathéodory-like decomposition result for semilinear sets:

Theorem 2.22 ([32]). *Let $k \geq 0$ and S be a semilinear subset of $\mathbb{N}^k$. There are a finite set I and for each $i \in I$ a vector $\vec{c}_i \in \mathbb{N}^k$ and a subset $P_i \subseteq \mathbb{N}^k$ which is linearly independent as a subset of the k-dimensional Euclidean space such that $S = \cup_{i \in I} \mathsf{L}(\vec{c}_i, P_i)$.*

This result was improved independently by Eilenberg and Schützenberger [24] and Ito [43]: they showed that the linear sets in Theorem 2.22 can be chosen pairwise disjoint. Clearly the family of semilinear sets is closed under union, direct product, sum, and homomorphism.

Proposition 2.23. *The family of semilinear sets is closed under Kleene star.*

Proof. Let $k \geq 0$ and S be a semilinear subset of $\mathbb{N}^k$. So, there are a finite set I and for each $i \in I$ a $\vec{c_i} \in \mathbb{N}^k$ and a finite $P_i \subseteq \mathbb{N}^k$ such that $S = \cup_{i \in I} \mathsf{L}(\vec{c_i}, P_i)$. Consider the semilinear set

$$T = \bigcup_{J \in 2^I} \mathsf{L}\left(\sum_{j \in J} \vec{c_j}, \bigcup_{j \in J} (\{\vec{c_j}\} \cup P_j)\right),$$

which fulfils $S \subseteq T \subseteq_{\text{sub}} \mathbb{N}^k$. On the other hand, each element of T is a finite sum of elements from S. That gives us $T = S^{*(\mathbb{N}^k, +, \vec{0})}$. $\square$

Three closure properties that are not easy to see were proven by Ginsburg and Spanier:

Theorem 2.24 ([32]). *The family of semilinear sets is closed under intersection, complementation, and inverse homomorphism.*

Presburger arithmetic is the predicate logic of the integers with addition, see the original paper by Presburger [64] and the *Grundlagen der Mathematik* by Hilbert and Paul Bernays [38]. A *Presburger sentence* is a Presburger formula without free variables. Presburger arithmetic is decidable, as shown by Presburger:

Theorem 2.25 ([38, 64]). *It is decidable if a Presburger sentence is true.*

A *modified* Presburger formula is a Presburger formula over the non-negative intergers. Using Theorem 2.24, Ginsburg and Spanier showed that the semilinear sets are those subsets of $\mathbb{N}^k$ that are definable by modified Presburger formulas:

Theorem 2.26 ([34]). *Let $k \geq 0$. For a modified Presburger formula P with exactly k free variables the set $\{\vec{x} \in \mathbb{N}^k \mid P(\vec{x}) \text{ is true}\}$ is semilinear and a semilinear representation is effectively constructable from P. For a semilinear set $S \subseteq \mathbb{N}^k$ a modified Presburger formula P with exactly k free variables such that $S = \{\vec{x} \in \mathbb{N}^k \mid P(\vec{x}) \text{ is true}\}$ is effectively constructable from a semilinear representation of S.*

From McKnight's Theorem [52] we can directly deduce:

Corollary 2.27. *For $k \geq 0$ each recognizable subset of $\mathbb{N}^k$ is semilinear.* □

Example 2.28. In the proof of Proposition 2.10 we have seen that the linear set

$$L\left(\vec{0}, \{(1,1)\}\right) = \{\,(n,n) \mid n \in \mathbb{N}\,\} \subset \mathbb{N}^2$$

is not recognizable. On the other hand, for $k \geq 0$ each $S \subseteq \mathbb{N}^k$ with $2 \leq |S| < \infty$ is recognizable, but not linear.

The next result was shown by Ginsburg and Spanier:

Proposition 2.29 ([33]). *For a semilinear set S it is decidable whether S is recognizable.*

From Proposition 2.6 and the closure of the family of recognizable sets under union it follows:

Proposition 2.30. *Each semilinear subset of $\mathbb{N}$ is recognizable.* □

We call a language *semilinear* if its Parikh-image is semilinear. For a unary alphabet Σ the Parikh-mapping $\psi_\Sigma : \Sigma^* \to \mathbb{N}$ is a monoid isomorphism. We get:

Corollary 2.31. *Over a unary alphabet the semilinear languages are exactly the regular languages.* □

The recognizable and the semilinear subsets of $\mathbb{N}^k$ are of great importance in formal language theory because of the following facts. Clearly a permutation closed language is regular if and only if its Parikh-image is recognizable. Using regular expressions it is not hard to see that the semilinear sets are the Parikh-images of the regular languages. Since each regular expression can be effectively transformed to an equivalent finite automaton and *vice versa*, we even get: For each semilinear set S a finite automaton A such that the Parikh-image of $L(A)$ equals S is effectively constructable. For each finite automaton A a semilinear representation of the Parikh-image of $L(A)$ is effectively constructable. The seminal result of Parikh states that even all context-free languages are semilinear:

Theorem 2.32 ([61, 62])**.** *For each PDA A a semilinear representation of the Parikh-image of $L(A)$ is effectively constructable.*

It follows that for each PDA a Parikh-equivalent finite automaton is effectively constructable. Hence, over a unary alphabet for each PDA an equivalent finite automaton is effectively constructable. Notice that the language $\{\, a^{2^n} \mid n \geq 0 \,\}$, which we have seen to be accepted by a deterministic linear bounded automaton in Example 2.12, is not semilinear.

Example 2.33. It can be shown that the permutation closed language

$$\{\, w \in \{a, b, c\}^* \mid |w|_a = |w|_b = |w|_c \,\}$$

is not context free, by setting $z = a^n b^n c^n$ in Lemma 2.13. The Parikh-image of this language is the linear set $\mathsf{L}\big(\vec{0}, \{(1, 1, 1)\}\big) \subset \mathbb{N}^3$.

From the given observations it follows that a language is permutation closed and semilinear if and only if it is the permutation closure of a regular language. Since **CS** is effectively closed under the operation of permutation closure, we get:

Proposition 2.34. *For each permutation closed semilinear language a linear bounded automaton accepting the language can be effectively constructed.* $\qquad\Box$

An automaton model for the family of permutation closed semilinear languages is given in the next section.

2.7 Jumping Finite Automata

For a λ-MNFA $A = (Q, \Sigma, \delta, S, F)$ we define the binary *jumping relation*, which is symbolically denoted by $\curvearrowright_A$ or just $\curvearrowright$ if it is clear which λ-MNFA we are referring to, on the set of configurations of A as follows. For $p \in Q$, $a \in \Sigma \cup \{\lambda\}$, $q \in \delta(p, a)$, and $v, w \in \Sigma^*$ we have $(p, vaw) \curvearrowright_A (q, vw)$. The language accepted by A interpreted as a *jumping finite automaton* is

$$L_J(A) = \{\, w \in \Sigma^* \mid \exists s \in S, f \in F : (s, w) \curvearrowright_A^* (f, \lambda_\Sigma) \,\}.$$

It can easily be seen that $L_J(A) = \mathsf{perm}(L(A))$. So, a language L is accepted by a λ-MNFA interpreted as a jumping finite automaton if and only if L is the permutation closure of a regular language. Hence, for each λ-MNFA A there is a DFAwttf B such that $L_J(B) = L_J(A)$. The family of languages accepted by λ-MNFAs interpreted as jumping finite automata is denoted by **JFA**. As we have seen in the last section, this language family consists of the permutation closed semilinear languages.

Example 2.35. Let A be the DFA $A = (\{q_0, q_1, q_2, q_3\}, \{a, b\}, \delta, \{q_0\}, \{q_3\})$, where δ consists of the rules $q_0 b \to q_1$, $q_0 a \to q_2$, $q_2 b \to q_3$, and $q_3 a \to q_2$. The automaton A is depicted in Figure 2.1. It holds

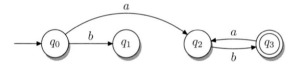

Figure 2.1: The DFA A with $L_J(A) = \{\, w \in \{a, b\}^* \mid |w|_a = |w|_b > 0 \,\}$.

$$L(A) = \{\, (ab)^n \mid n \geq 1 \,\} \subset \{a, b\}^*$$

and

$$L_J(A) = \mathsf{perm}(L(A)) = \{\, w \in \{a, b\}^* \mid |w|_a = |w|_b > 0 \,\}.$$

Since for each λ-MNFA A with input alphabet Σ a semilinear representation of the Parikh-image $\psi_\Sigma(L(A)) = \psi_\Sigma(L_J(A))$ is effectively constructable, for λ-MNFAs interpreted as jumping finite automata the word problem, emptiness, finiteness, universality, regularity, disjointness, inclusion, and equivalence are decidable. The following closure properties of **JFA** were given by Meduna and Zemek:

Proposition 2.36 ([53],[54]). *The family* **JFA** *is closed under union. On the other hand,* **JFA** *is not closed under intersection with regular languages, Kleene star, Kleene plus, concatenation, λ-free homomorphism, and substitution.*

For the operations intersection with regular languages, Kleene star, Kleene plus, concatenation, λ-free homomorphism, and substitution, Meduna and Zemek gave examples where the operand languages are in **JFA** but the resulting language is not even permutation closed. Meduna and Zemek claimed that **JFA** is also closed under intersection, complementation, and inverse homomorphism. Fernau *et al.* [29] noticed that the proofs for closure under intersection and under complementation by Meduna and Zemek do not work, but that **JFA** is anyway closed under intersection and under complementation because the family of semilinear sets is closed under these operations. In the following we show that the proof for closure under inverse homomorphism by Meduna and Zemek also does not work. However, the family of semilinear sets and the family of permutation closed languages are both closed under inverse homomorphism, so **JFA** is closed under inverse homomorphism, too.

In the book *Regulated Grammars and Automata* by Meduna and Zemek Chapter 17 is about jumping finite automata [54]. There, the authors use the same construction as for ordinary finite automata to show the closure of **JFA** under inverse homomorphism. It reads as follows. Let $A = (Q, \Gamma, \delta, \{s_0\}, F)$ be an NFA, Σ be an alphabet, and $h : \Sigma^* \to \Gamma^*$ be a monoid homomorphism. Consider the NFA $A' = (Q, \Sigma, \delta', \{s_0\}, F)$, where for all $p, q \in Q$ and $a \in \Sigma$ we have $q \in \delta'(p, a)$ if and only if $(p, h(a)) \vdash_A^* (q, \lambda_\Gamma)$. We get $L(A') = h^{-1}(L(A))$. Meduna and Zemek claim that it also holds $L_J(A') = h^{-1}(L_J(A))$. But this is not true in general as shown by the following counterexample.

Example 2.37. Consider the DFA A with input alphabet $\{a, b\}$ depicted on the left of Figure 2.2. It is easy to see that $L_J(A) = \{\, w \in \{a, b\}^* \mid |w|_a = |w|_b \,\}$. Define the monoid homomorphism $h : \{a\}^* \to \{a, b\}^*$ *via* $h(a) = ba$. Constructing A' as described above results in the DFA

Figure 2.2: (Left): The DFA A with $L_J(A) = \{\, w \in \{a, b\}^* \mid |w|_a = |w|_b \,\}$. (Right): The DFA A' induced from A and the monoid homomorphism $h : \{a\}^* \to \{a, b\}^*$ defined through $h(a) = ba$ by the standard construction on NFAs for the inverse homomorphism closure, fulfilling $L_J(A') = \{\lambda_{\{a\}}\}$.

drawn on the right of Figure 2.2. But then

$$L_J(A') = \{\lambda_{\{a\}}\} \neq \{a\}^* = h^{-1}(L_J(A)),$$

which shows the argument to be invalid.

Closure properties of the language families **CS**, **CF**, **DCF**, **REG**, and **JFA** are given in Table 2.1.

2.8 Right One-Way Jumping Finite Automata

For a DFA $A = (Q, \Sigma, \delta, \{s_0\}, F)$ we define the binary *right one-way jumping relation*, which is symbolically denoted by $\circlearrowright_A$ or just $\circlearrowright$ if it is clear which DFA we are referring to, on the set of configurations of A as follows. For $p \in Q$, $a \in \Sigma$, $q \in \delta(p, a)$, and $w \in \Sigma^*$ we have $(p, aw) \circlearrowright_A (q, w)$. On the other hand, for $p \in Q$, $a \in \Sigma$ with $\delta(p, a) = \emptyset$, and $w \in \Sigma^*$ we have $(p, aw) \circlearrowright_A (p, wa)$. The language accepted by A interpreted as a *right one-way jumping finite automaton* is

$$L_R(A) = \{\, w \in \Sigma^* \mid \exists f \in F : (s_0, w) \circlearrowright_A^* (f, \lambda_\Sigma) \,\}.$$

It is easy to see that

$$L(A) \subseteq L_R(A) \subseteq L_J(A) = \mathsf{perm}(L(A)).$$

	Language family				
Closed under	**CS**	**CF**	**DCF**	**REG**	**JFA**
Union	yes	yes	no	yes	yes
Intersection	yes	no	no	yes	yes
Intersection with reg. lang.	yes	yes	yes	yes	no
Complementation	yes	no	yes	yes	yes
Reversal	yes	yes	no	yes	yes
Kleene star	yes	yes	no	yes	no
Kleene plus	yes	yes	no	yes	no
Concatenation	yes	yes	no	yes	no
Homomorphism	no	yes	no	yes	no
λ-free homomorphism	yes	yes	no	yes	no
Inverse homomorphism	yes	yes	yes	yes	yes
Substitution	no	yes	no	yes	no
Permutation closure	yes	no	no	no	yes

Table 2.1: Closure properties of the language families **CS**, **CF**, **DCF**, **REG**, and **JFA**.

This implies that all these languages have the same Parikh-image, which is a semilinear set. If A is a DFAwttf, we get $L(A) = L_R(A)$. So, each regular language is accepted by a DFAwttf interpreted as a right one-way jumping finite automaton. The family of languages accepted by DFAs interpreted as right one-way jumping finite automata is denoted by **ROWJ**.

Example 2.38. We continue Example 2.35 and start with an example computation of the DFA A interpreted as a right one-way jumping finite automaton on the input word $aabbba$:

$$(q_0, aabbba) \circlearrowright (q_2, abbba) \circlearrowright^2 (q_3, bbaa) \circlearrowright^3 (q_2, abb)$$
$$\circlearrowright^2 (q_3, ba) \circlearrowright^2 (q_2, b) \circlearrowright (q_3, \lambda_{\{a,b\}})$$

That shows $aabbba \in L_R(A)$. Analogously, one can see that every word that contains the same number of a's and b's and that begins with an a is in $L_R(A)$. On the other hand, no other word can be accepted by A interpreted as a right one-way jumping finite automaton. Hence, we get

$$L_R(A) = \left\{ w \in \{a\} \cdot_{\{a,b\}} \{a,b\}^* \mid |w|_a = |w|_b \right\}.$$

Notice that this language is non-regular and not closed under permutation.

It is not hard to see that for each DFA interpreted as a right one-way jumping finite automaton an equivalent deterministic linear bounded automaton can effectively be constructed. Chigahara *et al.* showed that **ROWJ** is incomparable to **DCF** and to **CF** with the witness languages

$$\left\{ w \in \{a,b,c\}^* \mid |w|_a = |w|_b = |w|_c \right\}$$

and $\left\{ a^n b^n \mid n \geq 0 \right\} \subset \{a,b\}^*$ [14]. It was stated as an open problem by Chigahara *et al.* whether **JFA** $\subset$ **ROWJ** holds. Directly from the definitions it follows that each DFA interpreted as a right one-way jumping finite automaton can be understood as a deterministic right-revolving finite automaton as defined by Bensch *et al.* [11]. A sketch of a proof that there exists a language accepted by a deterministic right-revolving finite automaton which is not accepted by a DFA interpreted as a right one-way jumping finite automaton was given by Chigahara *et al.* They further showed the following closure properties of **ROWJ**.

Proposition 2.39 ([14]). *The family **ROWJ** is not closed under intersection, intersection with regular languages, reversal, Kleene star, Kleene plus, concatenation, concatenation with a regular language from the right, and substitution.*

Chigahara *et al.* conjectured that **ROWJ** is also not closed under union. For all $n \geq 0$ let **ROWJ$_n$** to be the family of all languages accepted by DFA s with at most n final states interpreted as right one-way jumping finite automata.

Part II
Semilinear Sets

3 Descriptional Complexity of Operations on Semilinear Sets

Since the family of semilinear sets is trivially closed under union, direct product, sum, homomorphism, and Kleene star, one very easily gets upper bounds for the parameters of the resulting semilinear set when one of these operations is applied to semilinear sets. Closure of the semilinear sets under intersection, complementation, and inverse homomorphism was shown by Ginsburg and Spanier [32]. However, for these operations no trivial proofs as for union, direct product, sum, homomorphism, and Kleene star are known. That is why the descriptional complexity of intersection, complementation, and inverse homomorphism on semilinear sets is harder to handle. In this chaper, we analyse the proofs for closure of the semilinear sets under intersection, complementation, and inverse homomorphism by Ginsburg and Spanier to derive upper bounds for the parameters of the resulting semilinear set when one of these operations is performed. We can notably improve the bound for complementation that we get from the work of Ginsburg and Spanier by additionally using a normal form result for semilinear sets by Anthony Widjaja To from 2010 [73]. Independently from us, Chistikov and Haase gave upper bounds for intersection and complementation in 2016 [15]. For intersection, their proof idea is also based on the proof of closure under intersection by Ginsburg and Spanier. For the upper bounds for complementation, Chistikov and Haase do not refer to the proof of closure under complementation by Ginsburg and Spanier, but use techniques from convex geometry. We compare the results of Chistikov and Haase to our results.

The following notion is necessary to understand the bounds for semilinear sets. For a representation through hybrid linear sets in $\mathbb{N}^k$, given as $R = (I, (C_i)_{i \in I}, (P_i)_{i \in I})$, we set

$$M(R) = \max \left(\{\, |C_i| \mid i \in I \,\} \right), \; m(R) = \max \left(\{\, ||\vec{c}||_\infty \mid \exists i \in I : \vec{c} \in C_i \,\} \right),$$
$$N(R) = \max \left(\{\, |P_i| \mid i \in I \,\} \right), \; n(R) = \max \left(\{\, ||\vec{p}||_\infty \mid \exists i \in I : \vec{p} \in P_i \,\} \right),$$
and
$$K(R) = |I|.$$

If we speak of the *parameters* of a representation R through hybrid linear sets, we mean the values $M(R)$, $m(R)$, $N(R)$, $n(R)$, and $K(R)$. For each representation R through hybrid linear sets in $\mathbb{N}^k$ we clearly have $M(R) \leq (m(R)+1)^k$ and $N(R) \leq (n(R)+1)^k$. Furthermore, by bunching together all constants that belong to the same set of periods, we get the following bound for the cardinality of the index set.

Proposition 3.1. *Let $k \geq 0$ and R be a representation through hybrid linear sets in $\mathbb{N}^k$. Then, there is a representation R' through hybrid linear sets in $\mathbb{N}^k$ with $S(R') = S(R)$, $m(R') = m(R)$, $n(R') \leq n(R)$, $N(R') \leq N(R)$,*

$$M(R') \leq \min \left(K(R) \cdot M(R), (m(R)+1)^k \right),$$

and

$$K(R') \leq \sum_{i=0}^{N(R)} \binom{(n(R)+1)^k}{i} \leq 2^{(n(R)+1)^k}.$$

$\square$

A rough summary of the bounds deduced by Chistikov and Haase and by us is given now. For two semilinear representations R_1 and R_2 in $\mathbb{N}^k$ there are upper bounds by Chistikov and Haase and by us for the parameters of a representation R through hybrid linear sets in $\mathbb{N}^k$ with $S(R) = S(R_1) \cap S(R_2)$. All these bounds are polynomial in the parameters of R_1 and R_2, where our bounds are more precise than the ones by Chistikov and Haase. For a semilinear representation R in $\mathbb{N}^k$ we give upper bounds for the parameters of a representation R' through hybrid linear sets in $\mathbb{N}^k$ with $S(R') = \mathbb{N}^k \setminus S(R)$ by analysing the proof by Ginsburg and Spanier that semilinear sets are closed under complementation. These bounds are double exponential in the parameters of R. By additionally applying a normal form result for semilinear sets by To, we get upper bounds that are only single exponential in the parameters of R. Chistikov and Haase gave a representation R'' through hybrid linear sets in $\mathbb{N}^k$ fulfilling $S(R'') = \mathbb{N}^k \setminus S(R)$ with upper bounds for $m(R'')$ and $n(R'')$, which are also single exponential in the parameters of R and better than our bounds. However, they did not give any bound for $K(R'')$. So, $K(R'')$ could possibly be double exponential in the parameters of R. For a semilinear representation R in $\mathbb{N}^{k_2}$ and a monoid homomorphism $h : \mathbb{N}^{k_1} \to \mathbb{N}^{k_2}$ we give upper bounds for the parameters of a representation R' through hybrid linear sets in $\mathbb{N}^{k_1}$ with $S(R') = h^{-1}(S(R))$. These bounds are polynomial in $||h||_{\max}$ and the parameters of R.

Most results on the descriptional complexity of operations on semilinear sets are based on a size estimate of minimal solutions of systems of linear equations. We use a result by Thiet-Dung Huynh [40], which is based on work by Joachim von zur Gathen and Malte Sieveking [75] and can be slightly generalized by a careful inspection of the original proof. The generalized result reads as follows:

Theorem 3.2 ([40], Theorem 2.6). *Let $n, m \geq 0$ be integers, $A \in \mathbb{Z}^{n \times m}$ be a matrix of rank r, and $\vec{b} \in \mathbb{Z}^n$ be a vector. Moreover, let Δ be the maximum of the absolute values of the $r \times r$ sub-determinants of the extended matrix $(A \mid \vec{b})$, and $M \subseteq \mathbb{N}^m$ be the set of minimal elements of $\{ \vec{x} \in \mathbb{N}^m \setminus \{\vec{0}\} \mid A\vec{x} = \vec{b} \}$. Then it holds $||M||_\infty \leq (m+1) \cdot \Delta$.*

We can estimate the value of the above mentioned sub-determinants with a corollary of Hadamard's inequality:

Theorem 3.3. *Let $r \geq 1$ be an integer, $A \in \mathbb{Z}^{r \times r}$ be a matrix, and m_i, for $1 \leq i \leq r$, be the maximum of the absolute values of the entries of the ith column of A. Then we have $|\det(A)| \leq r^{r/2} \Pi_{i=1}^r m_i$.*

3.1 Intersection

To get bounds for the intersection of semilinear sets, the following is done. We analyse the proof that semilinear sets are closed under intersection from Ginsburg and Spanier [32, Theorem 6.1]. Due to distributivity it suffices to look at the intersection of linear sets. Those coefficients of the periods of our linear sets, which deliver a vector in the intersection, are described by systems of linear equations. For the intersection of the linear sets we get a hybrid linear set, where the periods and constants are built out of the minimal solutions of these systems of equations. We estimate the size of the minimal solutions with the help of Theorems 3.2 and 3.3 in order to obtain upper bounds for the maximum norms of the resulting periods and constants.

Theorem 3.4. *Let $k \geq 1$ and R_1 and R_2 be semilinear representations in $\mathbb{N}^k$. We set $m = \max(m(R_1), m(R_2))$, $n = \max(n(R_1), n(R_2))$, and $N = \max(N(R_1), N(R_2))$. Then, there is a representation R through hybrid linear sets in $\mathbb{N}^k$ with $S(R) = S(R_1) \cap S(R_2)$, $K(R) = K(R_1) \cdot K(R_2)$,*

$$n(R) \leq 3k^{k/2}N^2 n^{k+1},$$

and

$$m(R) \leq 3k^{k/2}N^2 n^k \cdot \max(n, m) + m.$$

Proof. This proof is based on the proof by Ginsburg and Spanier that semilinear sets are closed under intersection [32, Theorem 6.1]. The theorem is trivial for $n = 0$, so let $n > 0$. Let $p, q \in \{1, 2, \ldots, N\}$ and

$$\vec{c}, \vec{d}, \vec{x_1}, \vec{x_2}, \ldots, \vec{x_p}, \vec{y_1}, \vec{y_2}, \ldots, \vec{y_q} \in \mathbb{N}^k$$

such that $||\vec{c}||_\infty \leq m$, $||\vec{d}||_\infty \leq m$, for all $i \in \{1, 2, \ldots, p\}$ we have $0 < ||\vec{x_i}||_\infty \leq n$, and for all $j \in \{1, 2, \ldots, q\}$ it holds $0 < ||\vec{y_j}||_\infty \leq n$. Denote by X and Y the subsets of $\mathbb{N}^{p+q}$ defined by

$$X = \left\{ (\lambda_1, \lambda_2, \ldots, \lambda_p, \mu_1, \mu_2, \ldots, \mu_q) \in \mathbb{N}^{p+q} \,\middle|\, \vec{c} + \sum_{i=1}^{p} \lambda_i \vec{x_i} = \vec{d} + \sum_{j=1}^{q} \mu_j \vec{y_j} \right\}$$

and

$$Y = \left\{ (\lambda_1, \lambda_2, \ldots, \lambda_p, \mu_1, \mu_2, \ldots, \mu_q) \in \mathbb{N}^{p+q} \ \middle| \ \sum_{i=1}^{p} \lambda_i \vec{x_i} = \sum_{j=1}^{q} \mu_j \vec{y_j} \right\}.$$

Let C and P be the sets of minimal elements of X and $Y \setminus \{\vec{0}\}$. Ginsburg and Spanier showed that $X = \mathsf{L}(C, P)$.

In order to estimate the size of $||C||_\infty$ and $||P||_\infty$ we use an alternative description of the vectors in X and Y in terms of matrix calculus. Let us define the matrix

$$H = (\vec{x_1} \mid \vec{x_2} \mid \cdots \mid \vec{x_p} \mid -\vec{y_1} \mid -\vec{y_2} \mid \cdots \mid -\vec{y_q}) \in \mathbb{Z}^{k \times (p+q)}.$$

Then it is easy to see that

$$\vec{x} \in X \quad \text{if and only if} \quad H\vec{x} = \vec{d} - \vec{c},$$

and

$$\vec{y} \in Y \quad \text{if and only if} \quad H\vec{y} = \vec{0}.$$

From Theorem 3.3 it follows that for each $r \in \{1, 2, \ldots, k\}$ the maximum of the absolute values of any $r \times r$ sub-determinant of the extended matrix $(H \mid \vec{0})$ is bounded from above by $k^{k/2} n^k$ because the maximum of the absolute values of the entries of the whole extended matrix $(H \mid \vec{0})$ is at most n. By Theorem 3.2 we conclude that

$$||P||_\infty \leq (p + q + 1) k^{k/2} n^k \leq 3k^{k/2} N n^k.$$

Analogously, for each $r \in \{1, 2, \ldots, k\}$ we can estimate the value of the maximum of the absolute values of any $r \times r$ sub-determinant of the extended matrix $(H \mid \vec{d} - \vec{c})$ by Theorem 3.3. For $m > 0$ it is bounded by $k^{k/2} n^{k-1} \cdot \max(n, m)$ because the maxima of the absolute values of the columns of $(H \mid \vec{d} - \vec{c})$ are bounded by n and m. Thus we have

$$||C||_\infty \leq (p + q + 1) k^{k/2} n^{k-1} \cdot \max(n, m) \leq 3k^{k/2} N n^{k-1} \cdot \max(n, m)$$

by Theorem 3.2.

Let $\tau : \mathbb{N}^{p+q} \to \mathbb{N}^k$ be the linear function given by

$$(\lambda_1, \lambda_2, \ldots, \lambda_p, \mu_1, \mu_2, \ldots, \mu_q) \mapsto \sum_{i=1}^{p} \lambda_i \vec{x_i}.$$

Then we have

$$\mathsf{L}\left(\vec{c}, \{\vec{x_1}, \vec{x_2}, \ldots, \vec{x_p}\}\right) \cap \mathsf{L}\left(\vec{d}, \{\vec{y_1}, \vec{y_2}, \ldots, \vec{y_q}\}\right) = \{\vec{c}\} + \tau(X).$$

The linearity of τ implies $\tau(X) = \mathsf{L}(\tau(C), \tau(P))$. So we have

$$\mathsf{L}\left(\vec{c}, \{\vec{x_1}, \vec{x_2}, \ldots, \vec{x_p}\}\right) \cap \mathsf{L}\left(\vec{d}, \{\vec{y_1}, \vec{y_2}, \ldots, \vec{y_q}\}\right) = \mathsf{L}(\{\vec{c}\} + \tau(C), \tau(P)).$$

It holds

$$||\tau(P)||_\infty \leq N \cdot ||P||_\infty \cdot n \leq 3k^{k/2}N^2n^{k+1}$$

and

$$||\tau(C)||_\infty \leq N \cdot ||C||_\infty \cdot n \leq 3k^{k/2}N^2n^k \cdot \max(n, m).$$

We get

$$||\{\vec{c}\} + \tau(C)||_\infty \leq m + ||\tau(C)||_\infty \leq 3k^{k/2}N^2n^k \cdot \max(n, m) + m.$$

The theorem follows with distributivity of intersection and union. □

If we assume k to be fixed and ℓ to be the maximum of m and n in Theorem 3.4, the maximum norms of the periods and constants in the representation through hybrid linear sets of the intersection are in $O(N^2\ell^{k+1} + \ell)$. So, all parameters of the representation through hybrid linear sets of the intersection are polynomial in the parameters of the semilinear representations of the operand sets. In this setting, Chistikov and Haase gave the upper bound $(N \cdot \max(\ell, 2))^{O(k)}$ for the maximum norms of the periods and constants and the same bound as we did for the cardinality of the index set in the representation through hybrid linear sets of the intersection [15, Theorem 6]. Hence, our bounds are more precise.

Since the set $\tau(P)$ in the proof of Theorem 3.4 does not depend on the constants $\vec{c}$ and $\vec{d}$ and in Theorem 3.4 it holds $N \leq (n+1)^k \leq (2n)^k$ if $n > 0$, we get:

Lemma 3.5. *Let $k \geq 1$ and R_1 and R_2 be representations through hybrid linear sets in $\mathbb{N}^k$. We set $m = \max(m(R_1), m(R_2))$ and $n =$*

$\max(n(R_1), n(R_2))$. *Then, there is a representation R through hybrid linear sets in $\mathbb{N}^k$ with $S(R) = S(R_1) \cap S(R_2)$, $K(R) = K(R_1) \cdot K(R_2)$,*

$$n(R) \le 4^{k+1}k^{k/2}n^{3k+1},$$

and

$$m(R) \le \max\left(4^{k+1}k^{k/2}n^{3k} \cdot \max(n,m), m\right).$$

$\square$

Now we turn to the intersection of more than two semilinear sets. The result is later utilized to explore the descriptional complexity of complementation.

Corollary 3.6. *Let $k \ge 1$ and the set X be finite. For every element $x \in X$ let R_x be a representations through hybrid linear sets in $\mathbb{N}^k$. We set $m = \max(\{\, m(R_x) \mid x \in X \,\})$ and $n = \max(\{\, n(R_x) \mid x \in X \,\})$ and assume $n > 0$. Then, there is a representation R through hybrid linear sets in $\mathbb{N}^k$ with $S(R) = \cap_{x \in X} S(R_x)$, $K(R) = \Pi_{x \in X} K(R_x)$,*

$$n(R) \le \left(4^{k+1}k^{k/2}n\right)^{(2\cdot|X|)^{\log(3k+1)}},$$

and

$$m(R) \le \left(4^{k+1}k^{k/2}n\right)^{(2\cdot|X|)^{\log(3k+1)}} \cdot \frac{\max(n,m)}{n}.$$

Proof. Set $p = \lceil\log(|X|)\rceil$. We show the following claim *via* induction over p.

Claim. There is a representation R through hybrid linear sets in $\mathbb{N}^k$ fulfilling $S(R) = \cap_{x \in X} S(R_x)$, $K(R) = \Pi_{x \in X} K(R_x)$,

$$n(R) \le \left(4^{k+1}k^{k/2}n\right)^{(3k+1)^p},$$

and

$$m(R) \le \left(4^{k+1}k^{k/2}n\right)^{(3k+1)^p} \cdot \frac{\max(n,m)}{n}.$$

Proof. The claim is clear for $p = 0$, so let $p > 0$. We build pairs of the elements in X. This gives us $\lfloor|X|/2\rfloor$ pairs of elements and an additional single element if $|X|$ is odd. Due to Lemma 3.5 we get for each such pair (x, y) a representation $R_{(x,y)}$ through hybrid linear sets in $\mathbb{N}^k$ with $S(R_{(x,y)}) = S(R_x) \cap S(R_y)$, $K(R_{(x,y)}) = K(R_x) \cdot K(R_y)$,

$$n(R_{(x,y)}) \leq 4^{k+1} k^{k/2} n^{3k+1},$$
$$m(R_{(x,y)}) \leq 4^{k+1} k^{k/2} n^{3k} \cdot \max(n, m).$$

Thus, we have a representation through hybrid linear sets for each of our pairs of elements from X and additionally a representation through hybrid linear sets for a single element out of X if $|X|$ is odd. If we now intersect the $\lceil |X|/2 \rceil$ represented semilinear sets, we get $\cap_{x \in X} S(R_x)$. Because of $\lceil \log (\lceil |X|/2 \rceil) \rceil = \lceil \log (|X|) \rceil - 1 = p - 1$, we can build this intersection by induction. This gives us a representation R through hybrid linear sets in $\mathbb{N}^k$ with $S(R) = \cap_{x \in X} S(R_x)$, $K(R) = \Pi_{x \in X} K(R_x)$,

$$n(R) \leq \left(4^{k+1} k^{k/2} \left(4^{k+1} k^{k/2} n^{3k+1} \right) \right)^{(3k+1)^{p-1}} \leq \left(4^{k+1} k^{k/2} n \right)^{(3k+1)^p},$$

and

$$m(R) \leq \left(4^{k+1} k^{k/2} \left(4^{k+1} k^{k/2} n^{3k+1} \right) \right)^{(3k+1)^{p-1}} \cdot \frac{4^{k+1} k^{k/2} n^{3k} \cdot \max(n, m)}{4^{k+1} k^{k/2} n^{3k+1}}$$
$$\leq \left(4^{k+1} k^{k/2} n \right)^{(3k+1)^p} \cdot \frac{\max(n, m)}{n}.$$

$\square$

The corollary follows directly from the claim. $\square$

3.2 Complementation

We analyse the proof by Ginsburg and Spanier that semilinear sets are closed under complementation [32, Lemmas 6.6, 6.7, 6.8, and 6.9] to get descriptional complexity results for the complement of a semilinear set. A linear set where the constant is the null-vector and the periods are linearly independent is complemented in Lemma 3.8. We continue by complementing a linear set with an arbitrary constant and linearly independent periods in Lemma 3.9. To complement a semilinear set where all the period sets are linearly independent in Lemma 3.10 we use DeMorgan's law: a semilinear set is a finite union of linear sets, so the complement is the intersection of the complements of the linear sets. For this intersection we use Corollary 3.6. Then we convert a linear set to a semilinear set with linearly independent period sets in Lemma 3.11. Finally we insert the bounds from Lemma 3.11 into the bounds from Lemma 3.10 to complement an arbitrary semilinear set in Theorem 3.12. First of all we complement a finite set, so we can exclude this case in the following results:

Lemma 3.7. *Let $k \geq 0$ and $F \subseteq \mathbb{N}^k$ be finite. Then, there is a representation R through hybrid linear sets in $\mathbb{N}^k$ fulfilling $S(R) = \mathbb{N}^k \setminus F$, $n(R) \leq 1$, $m(R) \leq ||F||_\infty + 1$, $N(R) = k$,*

$$M(R) = \max\left((||F||_\infty + 1)^k - |F|, k \right),$$

and $K(R) = 2$.

Proof. Let B_k be the standard basis of the k-dimensional Euclidean space. The set $\mathbb{N}^k \setminus F$ equals

$$\left\{ \vec{x} \in \mathbb{N}^k \mid ||\vec{x}||_\infty > ||F||_\infty \right\} \cup \left\{ \vec{x} \in \mathbb{N}^k \setminus F \mid ||\vec{x}||_\infty \leq ||F||_\infty \right\}$$
$$= \mathsf{L}\left(((||F||_\infty + 1) \cdot B_k, B_k) \cup \mathsf{L}\left(\left\{ \vec{x} \in \mathbb{N}^k \setminus F \mid ||\vec{x}||_\infty \leq ||F||_\infty \right\}, \emptyset \right).$$

$\square$

Next we consider the complement of a linear set where the constant is the null-vector and the periods are linearly independent.

Lemma 3.8. *Let $k \geq 1$ and $P \subset \mathbb{N}^k$ be non-empty and linearly independent as a subset of the k-dimensional Euclidean space. Then, there is a representation R through hybrid linear sets in $\mathbb{N}^k$ with $S(R) = \mathbb{N}^k \setminus \mathsf{L}(\vec{0}, P)$, $K(R) \leq 2^k + k - 1$, and*

$$\max(m(R), n(R)) \leq 3k^{k/2+1} (||P||_\infty)^k.$$

Proof. Let $p = |P|$ and $P = \{\vec{x_1}, \vec{x_2}, \ldots, \vec{x_p}\}$. By elementary vector space theory there exist $\vec{x_{p+1}}, \vec{x_{p+2}}, \ldots, \vec{x_k} \in \{\vec{e_{k,1}}, \vec{e_{k,2}}, \ldots, \vec{e_{k,k}}\}$ such that the set $\{\vec{x_1}, \vec{x_2}, \ldots, \vec{x_k}\}$ is a linearly independent subset of the k-dimensional Euclidean space. Let Δ be the absolute value of the determinant of the matrix $(\vec{x_1} \mid \vec{x_2} \mid \cdots \mid \vec{x_k})$. Moreover, let $\pi : \mathbb{N}^k \times \mathbb{N}^k \to \mathbb{N}^k$ be the projection on the first factor. For $J, K \subseteq \{1, 2, \ldots, k\}$ we define

$$A_J = \left\{ (\vec{y}, \vec{a}) \in \mathbb{N}^k \times \mathbb{N}^k \mid \forall j \in J : a_j > 0 \right\}$$

and

$$B_K = \left\{ (\vec{y}, \vec{a}) \in \mathbb{N}^k \times \mathbb{N}^k \mid \Delta \vec{y} + \sum_{i \in K} a_i \vec{x_i} = \sum_{i \in \{1, \ldots, k\} \setminus K} a_i \vec{x_i} \right\}.$$

Let Q_K and $D_{K,J}$ be the sets of minimal elements of $B_K \setminus \{\vec{0}\}$ and $B_K \cap A_J$. It holds $B_K \cap A_J = \mathsf{L}(D_{K,J}, Q_K)$. The linearity of π implies $\pi(B_K \cap A_J) = \mathsf{L}(\pi(D_{K,J}), \pi(Q_K))$. Next define the set B'_K to be

$$\left\{ (\vec{y}, \vec{a}) \in \mathbb{N}^k \times \mathbb{N}^k \mid \vec{y} + \sum_{i \in K} a_i \vec{x_i} = \sum_{i \in \{1, \ldots, k\} \setminus K} a_i \vec{x_i} \wedge \exists \vec{z} \in \mathbb{N}^k : \vec{y} = \Delta \vec{z} \right\}$$

and Q'_K and $D'_{K,J}$ to be the sets of minimal elements of $B'_K \setminus \{0\}$ and of $B'_K \cap A_J$. Then the mapping $f : B'_K \to B_K$ defined by $(\vec{y}, \vec{a}) \mapsto (\vec{y}/\Delta, \vec{a})$ is a bijection. The proof of the main result in the paper by von zur Gathen and Sieveking from 1978 [75] shows, together with Theorem 3.3, that

$$\max(||Q'_K||, ||D'_{K,J}||) \leq (2k\Delta+1)k^{k/2} \left(||P||_\infty\right)^k \leq \Delta \cdot (2k+1)k^{k/2} \left(||P||_\infty\right)^k.$$

With $Q_K = f(Q'_K)$ and $D_{K,J} = f(D'_{K,J})$ we get

$$\max\left(||\pi(Q_K)||, ||\pi(D_{K,J})||\right) \leq (2k+1)k^{k/2} \left(||P||_\infty\right)^k \leq 3k^{k/2+1} \left(||P||_\infty\right)^k.$$

Set $G_1 = \bigcup_{\emptyset \neq K \subseteq \{1, \ldots, k\}} \pi(B_K \cap A_K)$.

Let $\vec{y} \in \mathbb{N}^k$. Because $\{\vec{x_1}, \vec{x_2}, \ldots, \vec{x_k}\} \subset \mathbb{Q}^k$ is linearly independent, there are unique numbers $\lambda_{y,1}, \lambda_{y,2}, \ldots, \lambda_{y,k} \in \mathbb{Q}$ with $\vec{y} = \Sigma_{i=1}^k \lambda_{y,i} \vec{x_i}$. Then it holds $\vec{y} \in \mathsf{L}(\vec{0}, P)$ if and only if on the one hand for every $i \in \{1, 2, \ldots, p\}$ we have $\lambda_{y,i} \in \mathbb{N}$ and on the other hand for every $i \in \{p+1, p+2, \ldots, k\}$

it holds $\lambda_{y,i} = 0$. By Cramer's rule there are unique $\mu_{y,1}, \mu_{y,2}, \ldots, \mu_{y,k} \in \mathbb{Z}$ with $\Delta\vec{y} = \Sigma_{i=1}^{k} \mu_{y,i}\vec{x}_i$. Because of $\lambda_{y,i} = \mu_{y,i}/\Delta$, we get that $\vec{y} \in \mathsf{L}(\vec{0}, P)$ if and only if on the one hand for every $i \in \{1, 2, \ldots, p\}$ the number $\mu_{y,i}$ is a non-negative multiple of Δ and on the other hand for every $i \in \{p+1, p+2, \ldots, k\}$ it holds $\mu_{y,i} = 0$. The set G_1 consists of all $\vec{y} \in \mathbb{N}^k$ such that at least one of the $\mu_{y,i}$ is negative. This implies $G_1 \subseteq \mathbb{N}^k \setminus \mathsf{L}(\vec{0}, P)$.

Now we set $G_2 = \cup_{i=p+1}^{k} \pi(B_\emptyset \cap A_{\{i\}})$. This set consists of all $\vec{y} \in \mathbb{N}^k$ such that all the $\mu_{y,i}$ are non-negative and there exists $i \in \{p+1, p+2, \ldots, k\}$ such that $\mu_{y,i}$ is positive. This implies $G_2 \subseteq \mathbb{N}^k \setminus \mathsf{L}(\vec{0}, P)$.

For $i \in \{1, 2, \ldots, p\}$ and $r \in \{0, 1, \ldots, \Delta-1\}$ we set

$$E_{i,r} = \left\{ (\vec{y}, \vec{a}) \in \mathbb{N}^k \times \mathbb{N}^p \ \middle| \ \Delta\vec{y} = \sum_{j=1}^{p} a_j\vec{x}_j \text{ and } a_i \bmod \Delta = r \right\}.$$

Let $R_{i,r}$ be the set of minimal elements of $E_{i,r} \setminus \{\vec{0}\}$. For $r > 0$ it holds $E_{i,r} = \mathsf{L}(R_{i,r}, R_{i,0})$. We set $\pi' : \mathbb{N}^k \times \mathbb{N}^p \to \mathbb{N}^k$ to be the projection on the first factor. For $r > 0$ we get $\pi'(E_{i,r}) = \mathsf{L}(\pi'(R_{i,r}), \pi'(R_{i,0}))$. For every $i \in \{1, 2, \ldots, p\}$ it holds

$$\bigcup_{r=1}^{\Delta-1} \pi'(E_{i,r}) = \mathsf{L}\left(\bigcup_{r=1}^{\Delta-1} \pi'(R_{i,r}), \pi'(R_{i,0}) \right).$$

Let $i \in \{1, 2, \ldots, p\}$, $r \in \{0, 1, 2, \ldots, \Delta-1\}$, and $(\vec{y}, \vec{a}) \in R_{i,r}$. Then we have $||\vec{a}|| \leq \Delta$. This implies $||\vec{y}|| \leq p \cdot ||P||_\infty$. So we obtain that it holds $||\pi'(R_{i,r})|| \leq p \cdot ||P||_\infty$. Define $G_3 = \cup_{i=1}^{p} \cup_{r=1}^{\Delta-1} \pi'(E_{i,r})$. This is the set of all vectors $\vec{y} \in \mathbb{N}^k$ such that for all $j \in \{p+1, p+2, \ldots, k\}$ it holds $\mu_{y,j} = 0$, for all $j \in \{1, 2, \ldots, p\}$ we have $\mu_{y,j} \geq 0$, and for at least one $j \in \{1, 2, \ldots, p\}$ the number $\mu_{y,j}$ is not divisible by Δ. Thus we have $G_1 \cup G_2 \cup G_3 = \mathbb{N}^k \setminus \mathsf{L}(\vec{0}, P)$. $\qquad \square$

The following lemma gives a size estimation for the set $\mathbb{N}^k \setminus \mathsf{L}(\vec{x}, P)$, for an arbitrary vector $\vec{x}$, instead of the null-vector, as in the previous lemma, and a linearly independent P.

Lemma 3.9. *Let $k \geq 1$, $\vec{x} \in \mathbb{N}^k$, and $P \subset \mathbb{N}^k$ be non-empty and linearly independent as a subset of the k-dimensional Euclidean space.*

Then, there is a representation R through hybrid linear sets in $\mathbb{N}^k$ with $S(R) = \mathbb{N}^k \setminus \mathsf{L}(\vec{x}, P)$, $K(R) \leq 2^k + 2k - 1$,

$$n(R) \leq 3k^{k/2+1} \left(||P||_\infty\right)^k,$$
$$m(R) \leq 3k^{k/2+1} \left(||P||_\infty\right)^k + ||\vec{x}||_\infty,$$

and

$$M(R) \leq \max\left(4^k k^{k^2/2+k} \left(||P||_\infty\right)^{k^2}, ||\vec{x}||_\infty\right).$$

Proof. For $j \in \{1, 2, \ldots, k\}$ let

$$D_j = \left\{ \vec{y} \in \mathbb{N}^k \;\middle|\; y_j < x_j \wedge \forall \ell \in \{1, 2, \ldots, k\} \setminus \{j\} : y_\ell = 0 \right\}$$

and $Q_j = \{\overrightarrow{e_{k,1}}, \overrightarrow{e_{k,2}}, \ldots, \overrightarrow{e_{k,k}}\} \setminus \{\overrightarrow{e_{k,j}}\}$. Define $G = \cup_{j=1}^{k} \mathsf{L}(D_j, Q_j)$. This is the set of all $\vec{y} \in \mathbb{N}^k$ such that $\vec{x} \leq \vec{y}$ is false. So we have $G \subseteq \mathbb{N}^k \setminus \mathsf{L}(\vec{x}, P)$.

Now let $Y = \{ \vec{y} \in \mathbb{N}^k \mid \vec{x} \leq \vec{y} \}$. Then it holds $\mathbb{N}^k \setminus \mathsf{L}(\vec{x}, P) = G \cup (Y \setminus \mathsf{L}(\vec{x}, P))$, where $Y \setminus \mathsf{L}(\vec{x}, P) = (\mathbb{N}^k \setminus \mathsf{L}(\vec{0}, P)) + \{\vec{x}\}$. Due to Lemma 3.8 we have a representation $R' = (I, (C_i)_{i \in I}, (P_i)_{i \in I})$ through hybrid linear sets in $\mathbb{N}^k$ with $S(R') = \mathbb{N}^k \setminus \mathsf{L}(\vec{0}, P)$, $K(R') \leq 2^k + k - 1$, and $\max(m(R'), n(R')) \leq 3k^{k/2+1} \left(||P||_\infty\right)^k$. Then,

$$R'' = (I, (C_i + \{\vec{x}\})_{i \in I}, (P_i)_{i \in I})$$

is a representation through hybrid linear sets in $\mathbb{N}^k$ fulfilling $S(R'') = Y \setminus \mathsf{L}(\vec{x}, P)$. We obtain

$$M(R'') = M(R') \leq \left(3k^{k/2+1} \left(||P||_\infty\right)^k + 1\right)^k \leq \left(4k^{k/2+1}(||P||_\infty)^k\right)^k.$$

That proves the stated claim. $\qquad\qquad\qquad\qquad\qquad\qquad\qquad\square$

Now we are ready to deal with the complement of a semilinear set with linearly independent period sets.

Lemma 3.10. *Let $k \geq 1$ and $R = (I, (\vec{c}_i)_{i \in I}, (P_i)_{i \in I})$ be a semilinear representation in $\mathbb{N}^k$ such that $S(R)$ is infinite and for all $i \in I$ the set P_i is a linearly independent subset of the k-dimensional Euclidean* space. Then, there is a representation R' through hybrid linear sets in $\mathbb{N}^k$ with $S(R') = \mathbb{N}^k \setminus S(R)$, $K(R') = 2^{(k+1)K(R)}$,

$$n(R') \le (4kn(R))^{(k+2)(2K(R))^{\log(3k+1)}},$$

and

$$m(R') \le (4kn(R))^{(k+2)(2K(R))^{\log(3k+1)}} \cdot \max(m(R), 1).$$

Proof. Due to DeMorgan's law we have

$$\mathbb{N}^k \setminus S(R) = \mathbb{N}^k \setminus \bigcup_{i \in I} \mathsf{L}(\vec{c_i}, P_i) = \bigcap_{i \in I} \left(\mathbb{N}^k \setminus \mathsf{L}(\vec{c_i}, P_i) \right).$$

Because of Lemmas 3.7 and 3.9 for every $i \in I$ there is a representation R_i through hybrid linear sets in $\mathbb{N}^k$ with $S(R_i) = \mathbb{N}^k \setminus \mathsf{L}(\vec{c_i}, P_i)$, $K(R_i) \le 2^{k+1}$, $n(R_i) \le 3k^{k/2+1}(n(R))^k$, and $m(R_i) \le 3k^{k/2+1}(n(R))^k + m(R)$. Corollary 3.6 gives us a representation R' through hybrid linear sets in $\mathbb{N}^k$ with $S(R') = \cap_{i \in I} \left(\mathbb{N}^k \setminus \mathsf{L}(\vec{c_i}, P_i) \right)$, $K(R') = 2^{(k+1)K(R)}$,

$$n(R') \le \left(4^{k+1} k^{k/2} \cdot 3k^{k/2+1}(n(R))^k \right)^{(2K(R))^{\log(3k+1)}}$$
$$\le (4kn(R))^{(k+2)(2K(R))^{\log(3k+1)}},$$

and

$$m(R') \le \left(4^{k+1} k^{k/2} \cdot 3k^{k/2+1}(n(R))^k \right)^{(2K(R))^{\log(3k+1)}} \left(1 + \frac{m(R)}{3k^{k/2+1}(n(R))^k} \right)$$
$$\le (4kn(R))^{(k+2)(2K(R))^{\log(3k+1)}} \cdot \frac{3}{4} \left(1 + \frac{m(R)}{3} \right)$$
$$\le (4kn(R))^{(k+2)(2K(R))^{\log(3k+1)}} \cdot \max(m(R), 1).$$

This proves the stated claim. $\qquad\square$

Next we convert a linear set to a semilinear set with linearly independent period sets. The idea is the following: If the periods are linearly dependent we can rewrite our linear set as a semilinear set, where in each period set one of the original periods is removed. By doing this inductively the period sets get smaller and smaller until they are finally linearly independent.

Lemma 3.11. *Let $k \geq 1$, $\vec{x} \in \mathbb{N}^k$, and $P \subset \mathbb{N}^k$ be finite with $P\backslash\{\vec{0}\} \neq \emptyset$. Then, there exists a semilinear representation $R = (I, (\vec{c}_i)_{i \in I}, (P_i)_{i \in I})$ in $\mathbb{N}^k$ such that $S(R) = \mathsf{L}(\vec{x}, P)$,*

$$K(R) \leq \frac{|P|! \cdot (|P| + 1)!}{2^{|P|}} \left(k^{k/2} \left(||P||_\infty \right)^k \right)^{|P|-1},$$

$$m(R) \leq ||\vec{x}||_\infty + k^{k/2}/2 \cdot (|P| + 1)(|P| + 2) \left(||P||_\infty \right)^{k+1},$$

and for all $i \in I$ it holds $P_i \subseteq P$ and P_i is a linearly independent subset of the k-dimensional Euclidean space.

Proof. We prove this by induction on $|P|$. The statement of the lemma is clearly true for $|P| = 1$. So let $|P| \geq 2$ now. If P is linearly independent, the statement of the lemma is trivial. Thus we assume P to be linearly dependent. Then there exist $p \in \{1, 2, \ldots, \lfloor |P|/2 \rfloor\}$ and pairwise different vectors $\vec{x_1}, \vec{x_2}, \ldots, \vec{x_p}, \vec{y_1}, \vec{y_2}, \ldots, \overrightarrow{y_{|P|-p}} \in P$ such that $X = \{\vec{a} \in \mathbb{N}^{|P|} \setminus \{\vec{0}\} \mid H \cdot \vec{a} = \vec{0}\}$, where $H \in \mathbb{Z}^{k \times |P|}$ is the matrix given as

$$H = \left(\vec{x_1} \Big| \vec{x_2} \Big| \ldots \Big| \vec{x_p} \Big| - \vec{y_1} \Big| - \vec{y_2} \Big| \ldots \Big| - \overrightarrow{y_{|P|-p}} \right),$$

is not empty. Let $\vec{a}$ be a minimal element of X. From Theorems 3.2 and 3.3 we deduce $||\vec{a}||_\infty \leq (|P| + 1)k^{k/2} \left(||P||_\infty \right)^k$. For $j \in \{1, 2, \ldots, p\}$ let

$$C_j = \{\vec{x} + \lambda \vec{x_j} \mid \lambda \in \{0, 1, \ldots, a_j - 1\}\},$$

if $a_j > 0$, and $C_j = \{\vec{x}\}$, otherwise. Furthermore let $Q_j = P \backslash \{\vec{x_j}\}$. It was shown by Ginsburg and Spanier that $\cup_{j=1}^p \mathsf{L}(C_j, Q_j) = \mathsf{L}(\vec{x}, P)$ [32, proof of Lemma 6.6]. We can rewrite this set as $\cup_{j \in \{1,2,\ldots,p\}, \vec{c} \in C_j} \mathsf{L}(\vec{c}, Q_j)$. Here the cardinality of the index set is smaller than or equal to

$$|P|/2 \cdot ||\vec{a}||_\infty \leq k^{k/2}/2 \cdot |P| \cdot (|P| + 1) \left(||P||_\infty \right)^k$$

and for each such $\vec{c}$ we have

$$||\vec{c}||_\infty \leq ||\vec{x}||_\infty + ||\vec{a}||_\infty \cdot ||P||_\infty \leq ||\vec{x}||_\infty + k^{k/2}(|P| + 1) \left(||P||_\infty \right)^{k+1}.$$

By the induction hypothesis for each $j \in \{1, 2, \ldots, p\}$ and $\vec{c} \in C_j$, there exists a semilinear representation $R_{j,\vec{c}} = (I_{j,\vec{c}}, (\vec{c}_i)_{i \in I_{j,\vec{c}}}, (P_i)_{i \in I_{j,\vec{c}}})$ in $\mathbb{N}^k$ such that $S(R_{j,\vec{c}}) = \mathsf{L}(\vec{c}, Q_j)$,

$$K(R) \leq \frac{(|P| - 1)! \cdot |P|!}{2^{|P|-1}} \left(k^{k/2} \left(||P||_\infty \right)^k \right)^{|P|-2},$$

$$m(R)$$
$$\leq ||\vec{c}||_\infty + k^{k/2}/2 \cdot |P| \cdot (|P| + 1) \left(||P||_\infty\right)^{k+1}$$
$$\leq ||\vec{x}||_\infty + k^{k/2}(|P| + 1) \left(||P||_\infty\right)^{k+1} + k^{k/2}/2 \cdot |P| \cdot (|P| + 1) \left(||P||_\infty\right)^{k+1}$$
$$= ||\vec{x}||_\infty + k^{k/2}/2 \cdot (|P| + 1)(|P| + 2) \left(||P||_\infty\right)^{k+1},$$

and for all $i \in I_{j,\vec{c}}$ it holds $P_i \subseteq P$ and P_i is a linearly independent subset of the k-dimensional Euclidean space. This gives us

$$\bigcup_{j \in \{1,2,...,p\}, \vec{c} \in C_j, i \in I_{j,\vec{c}}} \mathsf{L}\left(\vec{c_i}, P_i\right) = \mathsf{L}(\vec{x}, P).$$

The cardinality of the index set is smaller than or equal to

$$k^{k/2}/2 \cdot |P| \cdot (|P| + 1) \left(||P||_\infty\right)^k \cdot \frac{(|P| - 1)! \cdot |P|!}{2^{|P|-1}} \left(k^{k/2} \left(||P||_\infty\right)^k\right)^{|P|-2}$$
$$= \frac{|P|! \cdot (|P| + 1)!}{2^{|P|}} \left(k^{k/2} \left(||P||_\infty\right)^k\right)^{|P|-1}.$$

$\square$

Now we can give the bounds for the complement of an arbitrary semilinear set that we deduce from the proof that semilinear sets are closed under complement by Ginsburg and Spanier.

Theorem 3.12. *Let $k \geq 1$ and R be a semilinear representation in $\mathbb{N}^k$. Then, there is a representation R' through hybrid linear sets in $\mathbb{N}^k$ with $S(R') = \mathbb{N}^k \setminus S(R)$,*

$$K(R') \leq 2^{121 \cdot K(R)\left(k^{k/2}(n(R)+1)^{3k}/11\right)^{N(R)+1}},$$
$$n(R') \leq 2^{\left(32 \cdot K(R)\left(k^{k/2}(n(R)+1)^{3k}/11\right)^{N(R)+1}\right)^{\log(3k+1)}},$$

and

$$m(R') \leq \max(m(R), 1) \cdot 2^{\left(121 \cdot K(R)\left(k^{k/2}(n(R)+1)^{3k}/11\right)^{N(R)+1}\right)^{\log(3k+1)}}.$$

Proof. The statement of the theorem is true if $S(R)$ is finite, because of Lemma 3.7. The statement is also true if $k = 1 = n(R)$, so let $kn(R) > 1$

in the following. By Lemmas 3.10 and 3.11 there is a representation R' through hybrid linear sets in $\mathbb{N}^k$ with $S(R') = \mathbb{N}^k \setminus S(R)$,

$$K(R') \leq 2^{(k+1)K(R)\frac{(N(R))!\cdot(N(R)+1)!}{2^{N(R)}}\left(k^{k/2}(n(R))^k\right)^{N(R)-1}},$$

$$n(R') \leq (4kn(R))^{(k+2)\left(K(R)\frac{(N(R))!\cdot(N(R)+1)!}{2^{N(R)-1}}\left(k^{k/2}(n(R))^k\right)^{N(R)-1}\right)^{\log(3k+1)}},$$

$$m(R') \leq (4kn(R))^{(k+2)\left(K(R)\frac{(N(R))!\cdot(N(R)+1)!}{2^{N(R)-1}}\left(k^{k/2}(n(R))^k\right)^{N(R)-1}\right)^{\log(3k+1)}}$$
$$\cdot \left(m(R) + k^{k/2}/2 \cdot (N(R)+1)(N(R)+2)\,(n(R))^{k+1}\right).$$

Stirling's formula gives us

$$(N(R))! \cdot (N(R)+1)!$$
$$< N(R)^{N(R)+1/2}e^{-N(R)+1}(N(R)+1)^{N(R)+3/2}e^{-N(R)}$$
$$= \left(\frac{N(R)}{N(R)+1}\right)^{N(R)+1/2}e^{-2N(R)+1}(N(R)+1)^{2(N(R)+1)}$$
$$= \left(1+(N(R))^{-1}\right)^{-(N(R)+1)+1/2}e^{-2N(R)+1}(N(R)+1)^{2(N(R)+1)}$$
$$< \sqrt{2}e^{-2N(R)}(N(R)+1)^{2(N(R)+1)}.$$

With

$$(N(R)+1)^2(n(R))^k \leq ((n(R)+1)^k+1)^2(n(R))^k$$
$$= \left((n(R)+1)^{2k}+2(n(R)+1)^k+1\right)(n(R))^k < \frac{4}{3}(n(R)+1)^{2k}(n(R)+1)^k$$

we have

$$K(R)\frac{(N(R))! \cdot (N(R)+1)!}{2^{N(R)}}\left(k^{k/2}\,(n(R))^k\right)^{N(R)-1}$$
$$< 2\sqrt{2}e^2K(R)\left((3e^2/2)^{-1}k^{k/2}\,(n(R)+1)^{3k}\right)^{N(R)+1} \cdot k^{-k}(n(R))^{-2k}.$$

From $4(k+1) \leq 3k^k(n(R))^{2k}$, $\frac{3}{2}\sqrt{2}e^2 < 16$, and $3e^2/2 > 11$ it follows

$$K(R') < 2^{16 \cdot K(R)\left(k^{k/2}(n(R)+1)^{3k}/11\right)^{N(R)+1}}.$$

Because of

$$4^{\log(3k+1)}(k+2) \cdot \log\left(4kn(R)\right) \le 3^{\log(3k+1)}k^{k\cdot\log(3k+1)}(n(R))^{2k\cdot\log(3k+1)}$$

and $3\sqrt{2}e^2 < 32$ it holds

$$n(R') < 2^{\left(32\cdot K(R)\left(k^{k/2}(n(R)+1)^{3k}/11\right)^{N(R)+1}\right)^{\log(3k+1)}}.$$

We have

$$1 + k^{k/2}/2 \cdot (N(R)+1)(N(R)+2)\,(n(R))^{k+1}$$
$$\le k^{k/2}(N(R)+1)^2\,(n(R))^{k+1} < \frac{4}{3}k^{k/2}\,(n(R)+1)^{3k+1}$$

and

$$\sqrt{\log\left(\frac{4}{3}k^{k/2}\,(n(R)+1)^{3k+1}\right)} < \frac{1}{2}\left(k^{k/2}\,(n(R)+1)^{3k}/11\right)^2.$$

With $3\sqrt{2}e^2 < 31.5$ we get

$$m(R') < \max(m(R),1) \cdot 2^{\left(32\cdot K(R)\left(k^{k/2}(n(R)+1)^{3k}/11\right)^{N(R)+1}\right)^{\log(3k+1)}}.$$

$\square$

Theorem 3.12 and Proposition 3.1 give us:

Corollary 3.13. *Let $k \ge 1$ be a fixed number, R be a semilinear representation in $\mathbb{N}^k$, and $\ell = \max(m(R), n(R))$. Then, there is a representation R' through hybrid linear sets in $\mathbb{N}^k$ with $S(R') = \mathbb{N}^k \setminus S(R)$,*

$$K(R') \in 2^{(O(\ell))^{4k(N(R)+1)}},$$

and

$$m(R'), n(R') \in 2^{(O(\ell))^{4k\cdot\log(3k+1)(N(R)+1)}}.$$

Proof. Let $f : \mathbb{N} \to \mathbb{N}$ be given by $x \mapsto (x+1)^k$. By Proposition 3.1 there is a representation R'' through hybrid linear sets in $\mathbb{N}^k$ with $S(R'') =$

$S(R), \max(m(R''), n(R'')) \leq \ell, N(R'') \leq N(R),$ and $K(R'') \leq \Sigma_{i=0}^{N(R)} \binom{f(\ell)}{i}$.
For $N(R) \leq 2$ we have $\Sigma_{i=0}^{N(R)} \binom{f(\ell)}{i} \leq 2(f(\ell))^{N(R)}$. For $N(R) \geq 3$ it holds

$$\sum_{i=0}^{N(R)} \binom{f(\ell)}{i} < \sum_{i=0}^{N(R)} \left(\frac{f(\ell)}{N(R)}\right)^i \frac{(N(R))^i}{i!} < \left(\frac{f(\ell)}{N(R)}\right)^{N(R)} e^{N(R)} < (f(\ell))^{N(R)}.$$

Thus, there exists a semilinear representation R''' in $\mathbb{N}^k$ fulfilling $S(R''') = S(R), \max(m(R'''), n(R''')) \leq \ell, N(R''') \leq N(R),$ and

$$K(R''') \leq K(R'')M(R'') \leq 2f(\ell)^{N(R)}(\ell+1)^k = 2(\ell+1)^{k(N(R)+1)}.$$

By Theorem 3.12 there is a representation R' through hybrid linear sets in $\mathbb{N}^k$ fulfilling $S(R') = \mathbb{N}^k \setminus S(R)$,

$$K(R') \in 2^{K(R''')(O(\ell))^{3k(N(R)+1)}},$$

and

$$m(R'), n(R') \in 2^{\left(K(R''')(O(\ell))^{3k(N(R)+1)}\right)^{\log(3k+1)}}.$$

$\square$

In Corollary 3.13 all parameters of the representation through hybrid linear sets of the complement are single exponential in the maximum norms of the periods and constants and double exponential in the maximum cardinality of the period sets of the semilinear representation of the operand set. We can improve these bounds, that we got from the proof that semilinear sets are closed under complementation by Ginsburg and Spanier, by additionally using the following normal form result for semilinear sets by To.

Theorem 3.14 ([73], Theorem 3.1). *Let $k \geq 1$, $\vec{x} \in \mathbb{N}^k$, and the finite set $P \subset \mathbb{N}^k$ fulfil $P \setminus \{\vec{0}\} \neq \emptyset$. Then, there exists a semilinear representation $R = (I, (\vec{c_i})_{i \in I}, (P_i)_{i \in I})$ in $\mathbb{N}^k$ such that $S(R) = \mathsf{L}(\vec{x}, P)$,*

$$K(R) \in O\left(|P|^{2k} \left(k^2 \cdot ||P||_\infty\right)^{2k^2+3k}\right),$$

$$m(R) \in ||x||_\infty + O\left(|P| \cdot (k^2 \cdot ||P||_\infty)^{2k+3}\right),$$

and for all $i \in I$ it holds $P_i \subseteq P$ and P_i is a linearly independent subset of the k-dimensional Euclidean space.

In his theorem, To states that for all $i \in I$ we have $|P_i| \leq k$, but he does not mention that the sets P_i are linearly independent. However, by looking at To's proof, one can easily see that the sets P_i are indeed linearly independent. The bounds from Theorem 3.14 are polynomial in $|P|$, in $||P||_\infty$, and in $||x||_\infty$. In Lemma 3.11, that we deduced from the proof of Theorem 2.22 by Ginsburg and Spanier, the bound for $K(R)$ is exponential in $|P|$. This is the reason why the bounds from Theorem 3.12 are double exponential in $N(R)$. From Lemmas 3.7 and 3.10 and Theorem 3.14 we directly get:

Theorem 3.15. *Let $k \geq 1$ and R be a semilinear representation in $\mathbb{N}^k$. Then, there is a representation R' through hybrid linear sets in $\mathbb{N}^k$ with $S(R') = \mathbb{N}^k \setminus S(R)$,*

$$K(R') \in 2^{K(R)O\left(k^{4k^2+6k+1}(n(R)+1)^{4k^2+3k}\right)},$$

$$n(R') \in 2^{\left(K(R)O\left(k^{4k^2+6k}(n(R)+1)^{4k^2+3k+1}\right)\right)^{\log(3k+1)}},$$

and

$$m(R') \in \max(m(R),1) \cdot 2^{\left(K(R)O\left(k^{4k^2+6k}(n(R)+1)^{4k^2+3k+1}\right)\right)^{\log(3k+1)}}.$$

$\square$

If we assume k to be fixed and ℓ to be the maximum of $m(R)$ and $n(R)$ in Theorem 3.15, in the representation through hybrid linear sets of the complement the cardinality of the index set is in

$$2^{K(R)\cdot O\left(\ell^{4k^2+3k}\right)}$$

and the maximum norms of the periods and constants are in

$$2^{\left(K(R)\cdot O\left(\ell^{4k^2+3k+1}\right)\right)^{\log(3k+1)}}.$$

Thus, all parameters of the representation through hybrid linear sets of the complement are single exponential in the parameters of the semilinear representation of the operand set. In this setting, independently from us Chistikov and Haase gave a representation through hybrid linear sets of the complement, for which the maximum norms of the periods and constants are in $(\max(\ell, 2))^{O(K(R))}$, using techniques from convex geometry [15, Theorem 21]. However, they did not give any bound for the cardinality of the index set. With Proposition 3.1 one gets an upper bound, which is double exponential in $K(R)$, for the cardinality of the index set of their representation through hybrid linear sets of the complement.

3.3 Inverse Homomorphism

Finally, the descriptional complexity of inverse homomorphism is considered. We follow the lines of the proof of inverse homomorphism closure of semilinear sets by Ginsburg and Spanier [32]. Since inverse homomorphism commutes with union, we only need to look at linear sets. The vectors in the preimage of a linear set, with respect to a homomorphism, can be described by a system of linear equations. Now we use the same techniques as in the proof of Theorem 3.4: Out of the minimal solutions of the system of equations we can build periods and constants of a representation through hybrid linear sets of the preimage. With Theorems 3.2 and 3.3 we estimate the size of the minimal solutions to get upper bounds for the maximum norms of the resulting periods and constants.

Proposition 3.16. *Let $k_1 \geq 0$, $k_2 \geq 1$, R be a semilinear representation in $\mathbb{N}^{k_2}$, and let $h : \mathbb{N}^{k_1} \to \mathbb{N}^{k_2}$ be a monoid homomorphism. Then, there is a representation R' through hybrid linear sets in $\mathbb{N}^{k_1}$ with $S(R') = h^{-1}(S(R))$, $K(R') = K(R)$,*

$$n(R') \leq (k_1 + N(R) + 1) \left(\sqrt{k_2} \cdot \max\left(n(R), ||h||_{\max}, 1\right) \right)^{k_2},$$

and

$$m(R') \leq m(R)(k_1 + N(R) + 1) \left(\sqrt{k_2} \cdot \max\left(n(R), ||h||_{\max}\right) \right)^{k_2}.$$

Proof. The proposition is true for $||h||_{\max} = 0$. Hence, let $||h||_{\max} > 0$ from now on. Furthermore, let $H \in \mathbb{N}^{k_2 \times k_1}$ be the matrix

$$\left(h\left(\overrightarrow{e_{k_1,1}}\right) \mid h\left(\overrightarrow{e_{k_1,2}}\right) \mid \cdots \mid h\left(\overrightarrow{e_{k_1,k_1}}\right) \right).$$

Moreover, let $p \in \{0, 1, \ldots, N(R)\}$ and $\vec{c}, \vec{y_1}, \vec{y_2}, \ldots, \vec{y_p} \in \mathbb{N}^{k_2}$ such that it holds $||\vec{c}||_\infty \leq m(R)$ and for all $i \in \{1, 2, \ldots, p\}$ we have $||\vec{y_i}||_\infty \leq n(R)$.

It holds

$$h^{-1}\left(\mathsf{L}\left(\vec{c}, \{\vec{y_1}, \vec{y_2}, \ldots, \vec{y_p}\}\right)\right)$$
$$= \left\{ \vec{x} \in \mathbb{N}^{k_1} \;\middle|\; \exists \lambda_1, \lambda_2, \ldots, \lambda_p \in \mathbb{N} : H\vec{x} = \vec{c} + \sum_{i=1}^{p} \lambda_i \vec{y_i} \right\}.$$

Now let $\tau : \mathbb{N}^{k_1} \times \mathbb{N}^p \to \mathbb{N}^{k_1}$ be the projection on the first component and let the matrix $J \in \mathbb{Z}^{k_2 \times (k_1+p)}$ be defined as $J = (H \mid -\vec{y_1} \mid -\vec{y_2} \mid \cdots \mid -\vec{y_p})$. We obtain

$$h^{-1}\left(\mathsf{L}\left(\vec{c}, \{\vec{y_1}, \vec{y_2}, \ldots, \vec{y_p}\}\right)\right) = \tau\left(\left\{ \vec{x} \in \mathbb{N}^{k_1+p} \;\middle|\; J\vec{x} = \vec{c} \right\}\right).$$

Let $C \subset \mathbb{N}^{k_1+p}$ be the set of minimal elements of $\left\{ \vec{x} \in \mathbb{N}^{k_1+p} \mid J\vec{x} = \vec{c} \right\}$ and $Q \subset \mathbb{N}^{k_1+p}$ be the set of minimal elements of the set $\left\{ \vec{x} \in \mathbb{N}^{k_1+p} \setminus \{\vec{0}\} \mid J\vec{x} = \vec{0} \right\}$. Then, we get the equation $\mathsf{L}(C, Q) = \left\{ \vec{x} \in \mathbb{N}^{k_1+p} \mid J\vec{x} = \vec{c} \right\}$. We derive from Theorems 3.2 and 3.3 that

$$||Q||_\infty \leq (k_1 + N(R) + 1)\left(\sqrt{k_2} \cdot \max\left(n(R), ||h||_{\max}\right)\right)^{k_2}$$
$$||C||_\infty \leq m(R)(k_1 + N(R) + 1)\left(\sqrt{k_2} \cdot \max\left(n(R), ||h||_{\max}\right)\right)^{k_2}.$$

Since τ is linear, it holds

$$\mathsf{L}(\tau(C), \tau(Q)) = h^{-1}\left(\mathsf{L}\left(\vec{c}, \{\vec{y_1}, \vec{y_2}, \ldots, \vec{y_p}\}\right)\right).$$

Moreover, we have $||\tau(Q)||_\infty \leq ||Q||_\infty$ and $||\tau(C)||_\infty \leq ||C||_\infty$. Because inverse homomorphism commutes with union, the proposition follows. $\square$

All parameters of the representation through hybrid linear sets of the preimage are polynomial in $||h||_{\max}$ and in the parameters of the semilinear representation of the operand set.

4 Semirecognizable Sets

McKnight's Theorem tells us that each recognizable subset of the free commutative monoid $\mathbb{N}^k$ is semilinear [52]. Because of Theorem 2.5 the recognizable subsets of $\mathbb{N}^k$ are well understood. In this chapter we define a generalization of recognizable subsets of monoids, called semirecognizable subsets, and especially study the semirecognizable subsets of $\mathbb{N}^k$, which turn out to build a subfamily of the semilinear sets. A subset S of a monoid M is called *semirecognizable* if there are a monoid N and a monoid homomorphism $f : M \to N$ such that $f(S)$ is finite and $S = f^{-1}(f(S))$. This is equivalent to the condition that the projection $S/\sim_S$ of S to its syntactic monoid is finite, by Proposition 2.2. If we replace "finite" by "a singleton" in the last two sentences, we get the notion of a *strongly semirecognizable* subset of a monoid. The term "semirecognizable" was chosen by us for the following reason. For a monoid M and an $S \subseteq M$ the syntactic monoid $M/\sim_S$ can be written as the disjoint union

$$M/\sim_S = \left\{ [a]_{\sim_S} \mid a \in S \right\} \cup \left\{ [a]_{\sim_S} \mid a \in M \setminus S \right\}.$$

The subset S is recognizable if $|M/\sim_S| < \infty$. That is why we call S semirecognizable if $|\{ [a]_{\sim_S} \mid a \in S \}| < \infty$. For the (strongly) semirecognizable subsets of $\mathbb{N}^k$ we show the following results. Each semirecognizable subset of $\mathbb{N}^k$ is a finite union of strongly semirecognizable linear subsets of $\mathbb{N}^k$. It is decidable whether a given semilinear set is (strongly) semirecognizable. A characterization when a linear subset of $\mathbb{N}^k$ is semirecognizable is given in terms of rational cones, which are a special type of convex cones. It was pointed out by Joseph Gubeladze and Mateusz Michałek [35] that "rational cones in $\mathbb{R}^d$ are important objects in toric algebraic geometry, combinatorial commutative algebra, geometric combinatorics, integer programming."

Lattices are introduced as special strongly semirecognizable subsets of $\mathbb{N}^k$. Their definition is inspired by the mathematical object of a lattice, which is of great importance in geometry and group theory, see *Sphere Packings, Lattices and Groups* by John H. Conway and Neil J. A. Sloane [21]. These lattices are subgroups of $\mathbb{R}^k$ that are isomorphic to $\mathbb{Z}^k$ and span the real vector space $\mathbb{R}^k$. Our lattices are defined like linear subsets of $\mathbb{N}^k$, but allowing integer coefficients for the period vectors, instead of only natural numbers. However our lattices are still, per definition, subsets of $\mathbb{N}^k$. If we only consider vectors where all components are "large enough," each strongly semirecognizable subset of $\mathbb{N}^k$ equals a lattice. Motivated by Theorem 2.22, we investigate Carathéodory-like decompositions of lattices and arbitrary (strongly) semirecognizable subsets of $\mathbb{N}^k$ and once more get a connection to rational cones. That is why we study these objects in more detail and show that the set of vectors with only non-negative components in a linear subspace of dimension n of $\mathbb{R}^k$ spanned by a subset of $\mathbb{N}^k$ always forms a rational cone spanned by a linearly independent subset of $\mathbb{N}^k$ if and only if $n \in \{0, 1, 2, k\}$. We state a characterization in terms of lattices when an arbitrary subset of $\mathbb{N}^k$ is a finite union of semirecognizable subsets of $\mathbb{N}^k$.

Parikh-preimages of semirecognizable subsets of $\mathbb{N}^k$, which form a proper subclass of the permutation closed semilinear languages, are investigated. In 1979 Latteux showed that each permutation closed semilinear language over a binary alphabet is context free [48]. Unfortunately the whole paper is written in French, which is why we were not even able to get the idea of the proof. In 2003 Rigo gave an easier proof for this interesting result in English [66]. He used a technique relying on factorization of words and constructed a context-free grammar generating the language in question. We provide an even simpler proof of the same result constructing a counter automaton accepting the considered language. Then, we study when permutation closed semirecognizable languages are regular, when they are context free, when they are accepted by a counter automaton, when they are deterministic context free, when they are accepted by a λ-free deterministic pushdown automaton, when they are accepted by a deterministic counter automaton, and when they are accepted by a λ-free deterministic counter automaton. Characterizations for all these properties are given in terms of linear algebra.

Some examples concerning semirecognizability are given:

Example 4.1. Consider the language $L = \{\, w \in \{a, b\}^* \mid |w|_a \neq |w|_b \,\}$. In Example 2.16 it was shown that the complement of L is in **DCF**. Since **DCF** is closed under complementation, L is also in **DCF**. For numbers $n, m \geq 1$ with $n \neq m$ we have $a^n \nsim_L a^m$. Hence, L is not semirecognizable.

Example 4.2. Every non-empty subset of the language

$$L = \{a\} \cdot_{\{a,b\}} \{b\}^{*\left(\{a,b\}^* \cdot_{\{a,b\}}, \lambda\right)} \cdot_{\{a,b\}} \{a\} \subset \{a, b\}^*$$

is strongly semirecognizable. Because there are uncountable many subsets of L, there exist strongly semirecognizable languages over a binary alphabet that are not recursively enumerable.

Example 4.3. The linear set $S = \mathsf{L}(\vec{0}, \{(1, 0), (1, 1)\}) \subset \mathbb{N}^2$ is not semirecognizable because for $n, m \geq 0$ with $n \neq m$ we have $(n, 0) \nsim_S (m, 0)$.

Example 4.4. In the proof of Proposition 2.10 we have seen that $S = \mathsf{L}(\vec{0}, \{(1, 1)\}) \subset \mathbb{N}^2$ is not recognizable. However, S is strongly semirecognizable.

The next result implies that each semirecognizable subset of $\mathbb{N}^k$ is semilinear.

Proposition 4.5. *For $k \geq 0$ each semirecognizable subset of $\mathbb{N}^k$ is a finite union of strongly semirecognizable linear subsets of $\mathbb{N}^k$.*

Proof. Since each semirecognizable set is a finite union of strongly semirecognizable sets, it suffices to show the proposition for strongly semirecognizable sets. Thus, let $k \geq 0$ and $S \subseteq \mathbb{N}^k$ be strongly semirecognizable. Let M be the set of minimal elements of S and N be the set of minimal elements of $D = \{\, \vec{x} - \vec{y} \mid \vec{x}, \vec{y} \in S, \vec{x} \geq \vec{y} \,\} \setminus \{\vec{0}\}$. For each $\vec{y_0} \in S$ it holds $D = \{\, \vec{x} - \vec{y_0} \mid \vec{x} \in S, \vec{x} \geq \vec{y_0} \,\} \setminus \{\vec{0}\}$. So, for each $\vec{z} \in S$ the strongly semirecognizable set $\{\, \vec{x} \in S \mid \vec{x} \geq \vec{z} \,\}$ equals $\mathsf{L}(\vec{z}, N)$. This gives us $S = \mathsf{L}(M, N)$. $\qquad\square$

The converse of Proposition 4.5 is not true as the following example shows.

Example 4.6. The set

$$S = \mathsf{L}\left(\vec{0}, \{(1,0)\}\right) \cup \mathsf{L}\left(\vec{0}, \{(1,1)\}\right) \subset \mathbb{N}^2$$

is the union of two strongly semirecognizable linear subsets of $\mathbb{N}^2$. However, S is not semirecognizable, because for $n, m \geq 0$ with $n \neq m$ it holds $(n, 0) \nsim_S (m, 0)$.

For each $n \geq 0$ a subset S of a monoid M is called *n-semirecognizable* if there are a monoid N and a monoid homomorphism $f : M \to N$ such that $|f(S)| \leq n$ and $S = f^{-1}(f(S))$. This is equivalent to the condition that the projection $S/\sim_S$ of S to its syntactic monoid consists of at most n elements, by Proposition 2.2. Hence, the 1-semirecognizable subsets of a monoid are the strongly semirecognizable subsets and the empty set. A subset of a monoid is semirecognizable if there exists a natural number n such that this subset is n-semirecognizable.

Example 4.7. For each $n \geq 0$ let L_n be the finite language $\{\, a^m \mid m < n \,\} \subset \{a\}^*$. For all $n \geq 0$ and $p, q \in \{0, 1, \ldots, n - 1\}$ with $p \neq q$ we have $a^p \nsim_{L_n} a^q$, which gives us $|L_n/\sim_{L_n}| = n$. So, for each $n > 0$ the language is n-semirecognizable, but not $(n - 1)$-semirecognizable.

Example 4.8. Let $n > 0$ and L_n be the infinite language

$$\{\, a^m \mid m \bmod (n + 1) \neq n \,\} \subset \{a\}^*.$$

The equivalence classes of $\sim_{L_n}$ are $[a^0], [a^1], \ldots, [a^n]$ and those are pairwise different. So, L_n is regular and $|L_n/\sim_{L_n}| = n$.

Example 4.9. Let $n > 0$ and L_n be the permutation closed language

$$\{\, w \in \{a, b\}^* \mid |w|_a \leq |w|_b < |w|_a + n \,\}.$$

For $v, w \in \{a, b\}^*$ it holds $v \sim_{L_n} w$ if and only if $|v|_b - |v|_a = |w|_b - |w|_a$. Hence, we get $|L_n/\sim_{L_n}| = n$ and that L_n is not regular. Consider the λ-free deterministic counter automaton

$$A = (\{s_{-1}, s_0, \ldots, s_n\}, \{a, b\}, \{\bot, C\}, \delta, s_0, \bot, \{s_0, s_1, \ldots, s_{n-1}\}),$$

where the map δ is defined as follows. For $i \in \{-1, 0, \ldots, n-1\}$ it holds

$$\delta(s_i, b, \bot) = \{(s_{i+1}, \bot)\} \qquad \text{and} \qquad \delta(s_{i+1}, a, \bot) = \{(s_i, \bot)\}.$$

We have

$$\delta(s_n, b, \bot) = \{(s_n, C\bot)\} \qquad \text{and} \qquad \delta(s_{-1}, a, \bot) = \{(s_{-1}, C\bot)\}.$$

It holds

$$\delta(s_n, b, C) = \{(s_n, CC)\} \qquad \text{and} \qquad \delta(s_n, a, C) = \{(s_n, \lambda)\}.$$

Furthermore, we have

$$\delta(s_{-1}, b, C) = \{(s_{-1}, \lambda)\} \qquad \text{and} \qquad \delta(s_{-1}, a, C) = \{(s_{-1}, CC)\}.$$

For all arguments which are not covered by the last sentences the value is the empty set. It holds $L(A) = L_n$.

More examples in the style of the last three ones are given in Section 4.4.

Ginsburg and Spanier [33] showed that it is decidable if a semilinear set is recognizable using a method by Richard Laing and Jesse B. Wright [47]. Similarly, we get from Theorems 2.25 and 2.26:

Corollary 4.10. *Let $k, n \geq 0$ and R be a semilinear representation in $\mathbb{N}^k$. Then, it is decidable whether $S(R)$ is semirecognizable and whether $S(R)$ is n-semirecognizable.*

Proof. Theorem 2.26 implies that we can effectively construct a modified Presburger formula P with exactly k free variables such that $S(R) = \{\, \vec{x} \in \mathbb{N}^k \mid P(\vec{x}) \text{ is true} \,\}$. The set $S(R)$ is semirecognizable if and only if the following Presburger sentence is true.

$$\exists m \in \mathbb{N} : \forall \vec{x} \in \mathbb{N}^k : \left(P(\vec{x}) \Rightarrow \exists \vec{y} \in \mathbb{N}^k : \left(\bigwedge_{i=1}^{k} y_i \leq m \right.\right.$$

$$\left.\left. \wedge\ \forall \vec{z} \in \mathbb{N}^k : (P(\vec{x} + \vec{z}) \Leftrightarrow P(\vec{y} + \vec{z})) \right)\right)$$

The set $S(R)$ is n-semirecognizable if and only if the following Presburger sentence is true.

$$\exists M \in \mathbb{N}^{k \times n} : \forall \vec{x} \in \mathbb{N}^k : \left(P(\vec{x}) \Rightarrow \exists \vec{y} \in \mathbb{N}^k : \left(\left(\bigvee_{j=1}^{n} \bigwedge_{i=1}^{k} y_i = m_{i,j} \right) \right. \right.$$

$$\left. \left. \wedge \; \forall \vec{z} \in \mathbb{N}^k : (P(\vec{x} + \vec{z}) \Leftrightarrow P(\vec{y} + \vec{z})) \right) \right)$$

The corollary follows from Theorem 2.25. $\qquad\qquad\qquad\square$

Closure properties of semirecognizable sets are considered. Let **SREC** be the family of semirecognizable subsets of monoids.

Proposition 4.11. *The family* **SREC** *is closed under intersection, direct product, and inverse homomorphism.*

Proof. Let $(M, \cdot, e)$ be a monoid and $S, T \subseteq M$ be semirecognizable subsets. So, there exist monoids $(M_0, \cdot_0, e_0)$ and $(M_1, \cdot_1, e_1)$ and monoid homomorphisms $f : M \to M_0$ and $g : M \to M_1$ such that $f(S)$ and $g(T)$ are finite, $S = f^{-1}(f(S))$, and $T = g^{-1}(g(T))$. The map $f \times g : M \to M_0 \times M_1$, $a \mapsto (f(a), g(a))$ is a monoid homomorphism. We have

$$(f \times g)(S \cap T) = (f \times g)(M) \cap f(S) \times g(T)$$

and

$$(f \times g)^{-1} \left((f \times g)(M) \cap f(S) \times g(T) \right) = S \cap T.$$

Hence, **SREC** is closed under intersection. Let $(M_2, \cdot_2, e_2)$ be a monoid and $h : M_2 \to M$ be a monoid homomorphism. We get

$$(f \circ h) \left(h^{-1}(S) \right) = f(h(M_2) \cap S) = f(h(M_2)) \cap f(S)$$

and

$$(f \circ h)^{-1} \left(f(h(M_2)) \cap f(S) \right) = h^{-1} \left(f^{-1}(f(h(M_2))) \cap S \right)$$
$$= (f \circ h)^{-1}((f \circ h)(M_2)) \cap h^{-1}(S) = h^{-1}(S).$$

Therefore, **SREC** is closed under inverse homomorphism. Let $(M_3, \cdot_3, e_3)$ be a monoid and $U \subseteq M_3$ be a semirecognizable subset. The map $F : M \times M_3 \to M_0$, $(a, b) \mapsto f(a)$ is a monoid homomorphism. It holds $F(S \times M_3) = f(S)$ and $F^{-1}(f(S)) = S \times M_3$, which implies that $S \times M_3$ is a semirecognizable subset of $M \times M_3$. Analogously, $M \times U$ is a semirecognizable subset of $M \times M_3$. So, $S \times M_3 \cap M \times U = S \times U$ is semirecognizable, which shows that **SREC** is closed under direct product. $\qquad\square$

The proof of Proposition 4.11 reveals:

Proposition 4.12. *Let $n, m \in \mathbb{N}$, M_1 and M_2 be monoids, the function $h : M_1 \to M_2$ be a monoid homomorphism, $S \subseteq M_1$ and $T \subseteq M_2$ be n-semirecognizable subsets, and $U \subseteq M_1$ be an m-semirecognizable subset. Then, the subsets $S \cap U \subseteq M_1$ and $U \times T \subseteq M_1 \times M_2$ are mn-semirecognizable and $h^{-1}(T) \subseteq M_1$ is n-semirecognizable.* $\qquad\square$

Example 4.13. Let $n, m \in \mathbb{N}$, $S = \{0, 1, \dots, n-1\}$, and T be the set $\{0, 1, \dots, m-1\}$. The subsets $S \subset \mathbb{N}$ and $S \times \mathbb{N} \subset \mathbb{N}^2$ are n-semirecognizable while $T \subset \mathbb{N}$ and $\mathbb{N} \times T \subset \mathbb{N}^2$ are m-semirecognizable. We have $S \times \mathbb{N} \cap \mathbb{N} \times T = S \times T \subset \mathbb{N}^2$ and $|(S \times T)/\!\sim_{S \times T}| = nm$. That shows that the bounds given in Proposition 4.12 are tight.

Example 4.14. Let $n, m, p, q \in \mathbb{N}_{>0}$ be numbers such that p and q are coprime, $n < p$, and $m < q$. Set $S = \{ i \in \mathbb{N} \mid i \bmod p < n \}$ and $T = \{ i \in \mathbb{N} \mid i \bmod q < m \}$. We get that $|S/\!\sim_S| = n$ and also that $|T/\!\sim_T| = m$. For $i, j \in \mathbb{N}$ it holds $i \sim_{S \cap T} j$ if and only if $i \bmod p = j \bmod p$ and $i \bmod q = j \bmod q$. So, we have $|S \cap T/\!\sim_{S \cap T}| = nm$.

Proposition 4.15. *There are subsets S, T, U, V, W, X, Y of $\mathbb{N}^2$ which are strongly semirecognizable and a monoid homomorphism $h : \mathbb{N}^2 \to \mathbb{N}^2$ such that $S \cup T$, $\mathbb{N}^2 \setminus U$, $V^{*(\mathbb{N}^2, +, \vec{0})}$, $W + X$, and $h(Y)$ are all nonsemirecognizable subsets of $\mathbb{N}^2$.*

Proof. Set

$$S = W = \{ (n, 0) \mid n \in \mathbb{N} \} \subset \mathbb{N}^2, \; T = U = X = \{ (n, n) \mid n \in \mathbb{N} \} \subset \mathbb{N}^2,$$

$$V = \{(2, 0), (1, 1)\} \subset \mathbb{N}^2, \text{ and } Y = \mathbb{N}^2. \text{ Clearly, all these subsets of } \mathbb{N}^2 \text{ are}$$

strongly semirecognizable. For $n, m \in \mathbb{N}$ with $n \neq m$ we have $(n, 0) \not\sim_{S \cup T}$

$(m, 0)$, $(n, 0) \sim_{\mathbb{N}^2 \setminus U} (m, 0)$, $(n, 0) \sim_{V^*(\mathbb{N}^2, +, \vec{0})} (m, 0)$, and $(n, 0) \sim_{W+X}$ $(m, 0)$. Thus, $S \cup T$, $\mathbb{N}^2 \setminus U$, $V^{*(\mathbb{N}^2, +, \vec{0})}$, and $W + X$ are non-semirecognizable subsets of $\mathbb{N}^2$. Let $h : \mathbb{N}^2 \to \mathbb{N}^2$ be the monoid homomorphism given by $h(a, b) = (a + b, b)$. It holds $h(Y) = W + X$. □

4.1 Rational Cones and Semirecognizability of Linear Sets

The semirecognizability of linear subsets of $\mathbb{N}^k$ is investigated. By the following straightforward observation the semirecognizability of a linear set does not depend on its constant vector.

Proposition 4.16. *Let $k \geq 0$, $\vec{c}, \vec{d} \in \mathbb{N}^k$, and $P \subseteq \mathbb{N}^k$ be finite. Then, there is a bijection between $\mathsf{L}(\vec{c}, P)/{\sim}_{\mathsf{L}(\vec{c}, P)}$ and $\mathsf{L}(\vec{d}, P)/{\sim}_{\mathsf{L}(\vec{d}, P)}$.* □

We show a connection between the semirecognizability of linear subsets of $\mathbb{N}^k$ and rational cones, which are a special type of convex cones. For $k \geq 0$ and a finite $S \subseteq \mathbb{R}^k$ the *convex cone spanned by S* is

$$\mathsf{cone}(S) = \left\{ \sum_{\vec{s} \in S} \lambda_{\vec{s}} \cdot \vec{s} \;\middle|\; \forall \vec{s} \in S : \lambda_{\vec{s}} \in \mathbb{R}_{\geq 0} \right\} \subseteq \mathbb{R}^k.$$

We give a proof of the conical version of Carathéodory's Theorem from convex geometry.

Theorem 4.17. *For $k \geq 0$ and a finite $S \subseteq \mathbb{R}^k$ it holds*

$$\mathsf{cone}(S) = \bigcup_{\substack{T \subseteq S, \\ T \text{ is linearly independent}}} \mathsf{cone}(T).$$

Proof. It clearly holds

$$\bigcup_{\substack{T \subseteq S, \\ T \text{ is linearly independent}}} \mathsf{cone}(T) \subseteq \mathsf{cone}(S).$$

So, let $\vec{x} \in \mathsf{cone}(S)$ and $T \subseteq S$ be minimal with $\vec{x} \in \mathsf{cone}(T)$. Hence, for each $\vec{t} \in T$ there is a $\lambda_{\vec{t}} \in \mathbb{R}_{>0}$ such that $\vec{x} = \Sigma_{\vec{t} \in T} \lambda_{\vec{t}} \cdot \vec{t}$. Assume that T is linearly dependent. Thus, for each $\vec{t} \in T$ there is a $\mu_{\vec{t}} \in \mathbb{R}$ such that $\vec{0} = \Sigma_{\vec{t} \in T} \mu_{\vec{t}} \cdot \vec{t}$ and for at least one $\vec{t} \in T$ it holds $\mu_{\vec{t}} > 0$. Set

$$\alpha = \min \left(\left\{ \frac{\lambda_{\vec{t}}}{\mu_{\vec{t}}} \,\middle|\, \vec{t} \in T,\, \mu_{\vec{t}} > 0 \right\} \right).$$

For every $\vec{t} \in T$ we have $\lambda_{\vec{t}} - \alpha\mu_{\vec{t}} \geq 0$. There is a $\vec{t_0} \in T$ with $\mu_{\vec{t_0}} > 0$ and $\alpha = \frac{\lambda_{\vec{t_0}}}{\mu_{\vec{t_0}}}$, which gives us $\lambda_{\vec{t_0}} - \alpha\mu_{\vec{t_0}} = 0$. It holds

$$\sum_{\vec{t} \in T \setminus \{\vec{t_0}\}} (\lambda_{\vec{t}} - \alpha\mu_{\vec{t}})\,\vec{t} = \sum_{\vec{t} \in T} (\lambda_{\vec{t}} - \alpha\mu_{\vec{t}})\,\vec{t} = \sum_{\vec{t} \in T} \lambda_{\vec{t}} \cdot \vec{t} - \alpha \sum_{\vec{t} \in T} \mu_{\vec{t}} \cdot \vec{t} = \vec{x}.$$

This is a contradiction to the minimality of T. Hence, T is linearly independent. $\qquad\square$

A *rational cone* in $\mathbb{R}^k$ is a set of the form $\mathsf{cone}(S)$ for a finite $S \subseteq \mathbb{Z}^k$. A *linearly independent rational cone* in $\mathbb{R}^k$ is a set of the form $\mathsf{cone}(S)$ for an $S \subseteq \mathbb{Z}^k$, which is a linearly independent subset of $\mathbb{R}^k$. From Theorem 4.17 we directly get:

Corollary 4.18. *Each rational cone is a finite union of linearly independent rational cones.* $\qquad\square$

A linear subset of $\mathbb{N}^k$ with period set P is semirecognizable if and only if the rational cone spanned by P equals the set of vectors with only non-negative components in the linear subspace spanned by P:

Theorem 4.19. *Let $k \geq 0$ and $P \subseteq \mathbb{N}^k$ be finite. Then, $\mathsf{L}(\vec{0}, P)$ is semirecognizable if and only if $\mathsf{cone}(P) = \mathsf{span}(P) \cap (\mathbb{R}_{\geq 0})^k$.*

Proof. Let $P = \{\vec{p_1}, \vec{p_2}, \ldots, \vec{p_{|P|}}\}$. Assume $\mathsf{cone}(P) = \mathsf{span}(P) \cap (\mathbb{R}_{\geq 0})^k$. Let U be the finite set

$$U = \left\{ \sum_{i=1}^{|P|} \nu_i \vec{p_i} \,\middle|\, \forall i \in \{1, 2, \ldots, |P|\} : (\nu_i \in \mathbb{R} \wedge 0 \leq \nu_i < 1) \right\} \cap \mathbb{N}^k.$$

Let $\vec{u} \in U$ and

$$V_{\vec{u}} = \left\{ (\xi_1, \xi_2, \ldots, \xi_{|P|}) \in \mathbb{N}^{|P|} \;\middle|\; \vec{u} + \sum_{i=1}^{|P|} \xi_i \vec{p_i} \in \mathsf{L}\left(\vec{0}, P\right) \right\}.$$

For all $\vec{v} \in V_{\vec{u}}$ and $\vec{w} \in \mathbb{N}^{|P|}$ with $\vec{v} \leq \vec{w}$ we have $\vec{w} \in V_{\vec{u}}$. Let $W_{\vec{u}}$ be the set of minimal elements of $V_{\vec{u}}$. For $\vec{x} \in \mathbb{N}^{|P|}$ let $X_{\vec{u},\vec{x}}$ be the finite set

$$\Big\{ (\max(0, \xi_1 - x_1), \max(0, \xi_2 - x_2), \ldots, \max(0, \xi_{|P|} - x_{|P|}))$$
$$\mid (\xi_1, \xi_2, \ldots, \xi_{|P|}) \in W_{\vec{u}} \Big\}.$$

For $\vec{x}, \vec{y} \in \mathbb{N}^{|P|}$ we have $\vec{x} + \vec{y} \in V_{\vec{u}}$ if and only if there is a $\vec{z} \in X_{\vec{u},\vec{x}}$ with $\vec{z} \leq \vec{y}$. The set $\{ X_{\vec{u},\vec{x}} \mid \vec{x} \in \mathbb{N}^{|P|} \}$ is finite.

Let now $t \in \mathsf{L}(\vec{0}, P)$ and $\vec{t'} \in \mathbb{N}^k$. If $\vec{t'} \notin \mathsf{span}(P)$, we have $\vec{t} + \vec{t'} \notin \mathsf{L}(\vec{0}, P)$. From now on, let $\vec{t'} \in \mathsf{span}(P)$. There is an $\vec{x} \in \mathbb{N}^{|P|}$ with $\vec{t} = \Sigma_{i=1}^{|P|} x_i \vec{p_i}$. Because the equation $\mathsf{cone}(P) = \mathsf{span}(P) \cap (\mathbb{R}_{\geq 0})^k$ holds, there are $\vec{u'} \in U$ and $\vec{x'} \in \mathbb{N}^{|P|}$ such that $\vec{t'} = \vec{u'} + \Sigma_{i=1}^{|P|} x'_i \vec{p_i}$. We have $\vec{t} + \vec{t'} \in \mathsf{L}(\vec{0}, P)$ if and only if $\vec{x} + \vec{x'} \in V_{\vec{u'}}$, which holds if and only if there is a $\vec{z} \in X_{\vec{u'},\vec{x}}$ with $\vec{z} \leq \vec{x'}$. So, the set $\{ \vec{r} \in \mathbb{N}^k \mid \vec{t} + \vec{r} \in \mathsf{L}(\vec{0}, P) \}$ only depends on the map $U \to \{ X_{\vec{v},\vec{y}} \mid \vec{v} \in U, \vec{y} \in \mathbb{N}^{|P|} \}$ given by $\vec{v} \mapsto X_{\vec{v},\vec{x}}$. This shows that $\mathsf{L}(\vec{0}, P)$ is semirecognizable.

Assume now that $\mathsf{cone}(P) \neq \mathsf{span}(P) \cap (\mathbb{R}_{\geq 0})^k$, which means $\mathsf{cone}(P) \subset \mathsf{span}(P) \cap (\mathbb{R}_{\geq 0})^k$. So, let $\vec{x} \in \mathsf{span}(P) \cap (\mathbb{R}_{\geq 0})^k \setminus \mathsf{cone}(P)$ and $T = \{ i \in \{1, 2, \ldots, k\} \mid x_i = 0 \}$. By basic linear algebra there is a non-empty linearly independent $Q \subset \mathbb{Q}^k$ with

$$\mathsf{span}(Q) = \left\{ \vec{y} \in \mathsf{span}(P) \;\middle|\; \pi_{k,T}(\vec{y}) = \vec{0} \right\}.$$

Because $\vec{x} \in \mathsf{span}(Q) \cap (\mathbb{R}_{\geq 0})^k \setminus \mathsf{cone}(P)$ and $\mathsf{cone}(P)$ is a topological closed set, see the paper *Convex cones in finite-dimensional real vector spaces* by Milan Studený [71], there is a

$$\vec{z} \in \mathsf{span}(Q) \cap (\mathbb{Q}_{\geq 0})^k \setminus \mathsf{cone}(P) \subseteq \mathsf{span}(P) \cap (\mathbb{Q}_{\geq 0})^k \setminus \mathsf{cone}(P).$$

So, there are $\lambda_1, \mu_1, \lambda_2, \mu_2, \ldots, \lambda_{|P|}, \mu_{|P|} \in \mathbb{Q}_{\geq 0}$ with

$$\sum_{i=1}^{|P|} (\lambda_i - \mu_i)\vec{p_i} \in (\mathbb{Q}_{\geq 0})^k \setminus \mathsf{cone}(P).$$

Multiplying with the lowest common denominator we may even assume that the values $\lambda_1, \mu_1, \lambda_2, \mu_2, \ldots, \lambda_{|P|}, \mu_{|P|}$ are in $\mathbb{N}$. Let

$$R = \min\left(\left\{ r \in \mathbb{R} \;\middle|\; \sum_{i=1}^{|P|} (\lambda_i + r\mu_i)\vec{p_i} \in \mathsf{cone}(P) \right\}\right) \in (-1, 0].$$

The set of minimal elements of $\mathbb{N}^k \cap \mathsf{span}(P) \setminus \{\vec{0}\}$ is denoted by M. We have $\mathbb{N}^k \cap \mathsf{span}(P) = \mathsf{L}(\vec{0}, M)$. For each $B \subseteq P$ which is a linearly independent subset of $\mathbb{R}^k$ with $\mathsf{span}(B) = \mathsf{span}(P)$ let N_B be the lowest common multiply of

$$\left\{ \min\left(\left\{ n \in \mathbb{N} \setminus \{0\} \;\middle|\; \exists(\kappa : B \to \mathbb{Z}) : n\vec{v} = \sum_{\vec{b} \in B} \kappa\left(\vec{b}\right) \cdot \vec{b} \right\}\right) \;\middle|\; \vec{v} \in M \right\},$$

which gives us

$$\mathsf{L}\left(\vec{0}, B\right) \cap (N_B \cdot \mathbb{N}^k) = \mathsf{cone}(B) \cap (N_B \cdot \mathbb{N}^k).$$

Let N be the lowest common multiply of all the numbers N_B. By Theorem 4.17 the set $\mathsf{cone}(P)$ is the union of all the sets $\mathsf{cone}(B)$ where $B \subseteq P$ is a linearly independent subset of $\mathbb{R}^k$ with $\mathsf{span}(B) = \mathsf{span}(P)$. It follows

$$\mathsf{L}\left(\vec{0}, P\right) \cap (N \cdot \mathbb{N}^k) = \mathsf{cone}(P) \cap (N \cdot \mathbb{N}^k).$$

For all $\nu, \xi \in \mathbb{N}$ it holds

$$\nu N \cdot \left(\sum_{i=1}^{|P|} \mu_i \vec{p_i}\right) + \xi N \cdot \sum_{i=1}^{|P|} (\lambda_i - \mu_i)\vec{p_i} \in \mathsf{L}\left(\vec{0}, P\right)$$

if and only if $\nu \geq \xi \cdot (R + 1)$. So, for $\nu_1, \nu_2 \in \mathbb{N}$ with $\nu_1 \neq \nu_2$ we have

$$\left[\nu_1 N \cdot \left(\sum_{i=1}^{|P|} \mu_i \vec{p_i}\right) \right]_{\sim_{\mathsf{L}(\vec{0}, P)}} \neq \left[\nu_2 N \cdot \left(\sum_{i=1}^{|P|} \mu_i \vec{p_i}\right) \right]_{\sim_{\mathsf{L}(\vec{0}, P)}},$$

which implies that $\mathsf{L}(\vec{0}, P)$ is not semirecognizable. $\qquad\square$

For linear subsets of $\mathbb{N}^k$ with a linearly independent period set semirecognizability implies strong semirecognizability:

Corollary 4.20. *Let* $k \geq 0$ *and* $P \subseteq \mathbb{N}^k$ *be a linearly independent subset of* $\mathbb{R}^k$ *such that* $\mathsf{L}(\vec{0}, P)$ *is semirecognizable. Then,* $\mathsf{L}(\vec{0}, P)$ *is strongly semirecognizable.*

Proof. Let $P = \{\vec{p_1}, \vec{p_2}, \ldots, \vec{p_{|P|}}\}$, $\vec{x} \in \mathsf{L}(\vec{0}, P)$, and $\vec{y} \in \mathbb{N}^k$. If $\vec{y} \notin \mathsf{span}(P)$, we get $\vec{x} + \vec{y} \notin \mathsf{L}(\vec{0}, P)$. If $\vec{y} \in \mathsf{span}(P)$, we have $\vec{y} \in \mathsf{cone}(P)$ because of Theorem 4.19. So, there are uniquely determined $\lambda_1, \lambda_2, \ldots, \lambda_{|P|} \in \mathbb{R}_{\geq 0}$ with $\vec{y} = \Sigma_{i=1}^{|P|} \lambda_i \vec{p_i}$. It holds $\vec{x} + \vec{y} \in \mathsf{L}(\vec{0}, P)$ if and only if for each $i \in \{1, 2, \ldots, |P|\}$ we have $\lambda_i \in \mathbb{N}$. This proves that $\mathsf{L}(\vec{0}, P)$ is strongly semirecognizable. $\square$

The next example shows that Corollary 4.20 does not hold if we do not demand the period set to be linearly independent.

Example 4.21. Let $P = \{2, 3\} \subset \mathbb{N}$. We have $\mathsf{L}(0, P) = \mathbb{N} \setminus \{1\}$, which gives us $\left| \mathsf{L}(\vec{0}, P)/\sim_{\mathsf{L}(\vec{0}, P)} \right| = 2$. Hence, $\mathsf{L}(0, P)$ is recognizable, but not strongly semirecognizable.

We can easily check whether a linear subset of $\mathbb{N}^k$ with a linearly independent period set is semirecognizable:

Corollary 4.22. *Let* $k \geq 0$ *and* $P \subseteq \mathbb{N}^k$ *be a linearly independent subset of* $\mathbb{R}^k$. *Then,* $\mathsf{L}(\vec{0}, P)$ *is semirecognizable if and only if for all vectors* $\vec{p} \in P$ *there is an* $i \in \{1, 2, \ldots, k\}$ *such that* $p_i > 0$ *and for all* $\vec{q} \in P \setminus \{\vec{p}\}$ *it holds* $q_i = 0$.

Proof. Assume that for all $\vec{p} \in P$ there is an $i \in \{1, 2, \ldots, k\}$ such that $p_i > 0$ and for all $\vec{q} \in P \setminus \{\vec{p}\}$ it holds $q_i = 0$. Then, $\mathsf{L}(\vec{0}, P)$ is semirecognizable by Theorem 4.19. Assume now that $\mathsf{L}(\vec{0}, P)$ is semirecognizable. Then we have $\mathsf{cone}(P) = \mathsf{span}(P) \cap (\mathbb{R}_{\geq 0})^k$ by Theorem 4.19. Thus, for all $\vec{p} \in P$ and $\lambda \in \mathbb{R}_{>0}$ it holds $\Sigma (P \setminus \{\vec{p}\}) - \lambda\vec{p} \notin (\mathbb{R}_{\geq 0})^k$. It follows that for all $\vec{p} \in P$ there is an $i \in \{1, 2, \ldots, k\}$ such that $p_i > 0$ and for all $\vec{q} \in P \setminus \{\vec{p}\}$ it holds $q_i = 0$. $\square$

Hence, semirecognizability is passed on from a linear subsets of $\mathbb{N}^k$ with a linearly independent period set P to all linear subsets of $\mathbb{N}^k$ whose period sets are subsets of P:

Corollary 4.23. *Let $k \geq 0$, $P \subseteq \mathbb{N}^k$ be a linearly independent subset of $\mathbb{R}^k$ such that $\mathsf{L}(\vec{0}, P)$ is semirecognizable, and $Q \subseteq P$. Then, $\mathsf{L}(\vec{0}, Q)$ is also semirecognizable.* □

4.2 Rational Cones and Carathéodory-Like Decompositions of Lattices

We study special strongly semirecognizable subsets of $\mathbb{N}^k$, which we call lattices. For all $k \geq 0$ a subset of $\mathbb{N}^k$ is called a *lattice in $\mathbb{N}^k$* if it is of the form

$$\mathsf{La}(\vec{c}, P) = \left\{ \vec{c} + \sum_{\vec{p} \in P} \lambda_{\vec{p}} \cdot \vec{p} \,\middle|\, \forall \vec{p} \in P : \lambda_{\vec{p}} \in \mathbb{Z} \right\} \cap \mathbb{N}^k$$

for a finite set $P \subseteq \mathbb{N}^k$ and a $\vec{c} \in \mathbb{N}^k$, where P is called the set of *periods* and $\vec{c}$ is called the *constant vector* of this representation. One can can think of a lattice as extending "the pattern" of a linear set to the whole $\mathbb{N}^k$. For $\vec{c} \in \mathbb{N}^k$, a finite $P \subseteq \mathbb{N}^k$, $\vec{x} \in \mathsf{La}(\vec{c}, P)$, and $\vec{y} \in \mathbb{N}^k$ it holds $\vec{x} + \vec{y} \in \mathsf{La}(\vec{c}, P)$ if and only if $\vec{y} \in \mathsf{La}(\vec{0}, P)$. That gives us:

Proposition 4.24. *For $k \geq 0$ each lattice in $\mathbb{N}^k$ is a strongly semirecognizable subset of $\mathbb{N}^k$.* □

With Proposition 4.5 it follows that each lattice is a semilinear set. We give an example of a lattice:

Example 4.25. Consider the constant vector $\vec{c} = (4,4)$ and the period vectors $(1,2)$ and $(2,0)$. A graphical presentation of the linear set $\mathsf{L}(\vec{c}, P)$ with $P = \{(1,2), (2,0)\}$ is given on the left of Figure 4.1. The constant vector $\vec{c}$ is drawn as a dashed arrow and both periods are depicted as solid arrows. The dots indicate the elements that belong to $\mathsf{L}(\vec{c}, P)$. The lattice $\mathsf{La}(\vec{c}, P)$ is drawn in the middle of Figure 4.1. Again, the constant

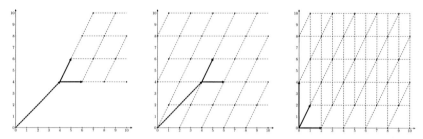

Figure 4.1: The linear set $\mathsf{L}(\vec{c}, P)$ with $\vec{c} = (4,4)$ and $P = \{(1,2),(2,0)\}$ drawn on the left. The black dots indicate membership in $\mathsf{L}(\vec{c}, P)$. The lattice $\mathsf{La}(\vec{c}, P)$ is depicted in the middle. Here the black dots refer to membership in $\mathsf{La}(\vec{c}, P)$. On the right a representation of $\mathsf{La}(\vec{c}, P)$ as a linear set is shown. The constant vector $\vec{0}$ is not shown and the period vectors are drawn as solid arrows.

vector is dashed, while both periods are solid arrows. Since now integer coefficients are allowed, there are new elements compared to $\mathsf{L}(\vec{c}, P)$ that belong to $\mathsf{La}(\vec{c}, P)$. On the right of Figure 4.1 it is shown that $\mathsf{La}(\vec{c}, P)$ can be written as a linear set by using the constant vector $\vec{0}$ and the three period vectors drawn as solid arrows, that is, $\mathsf{La}(\vec{c}, P) = \mathsf{L}(\vec{0}, \{(1,2),(2,0),(0,4)\})$.

For each lattice there is a linearly independent period set:

Proposition 4.26. *For $k \geq 0$ and a lattice L in $\mathbb{N}^k$ there are a vector $\vec{c} \in \mathbb{N}^k$ and a $P \subseteq \mathbb{N}^k$ which is a linearly independent subset of $\mathbb{R}^k$ such that $L = \mathsf{La}(\vec{c}, P)$.*

Proof. By definition there are a $\vec{c} \in \mathbb{N}^k$ and a finite $Q \subseteq \mathbb{N}^k$ with $L = \mathsf{La}(\vec{c}, Q)$. The proposition holds if $Q \subseteq \{\vec{0}\}$, so assume $Q \setminus \{\vec{0}\} \neq \emptyset$. Let G be the subgroup of $\mathbb{Z}^k$ that is generated by Q, d be the greatest common divisor of the components of the vector $\Sigma_{\vec{q} \in Q}\, \vec{q}$, and

$$m = \min\left(\left\{ a \in \mathbb{N}_{>0} \,\middle|\, \frac{a}{d} \cdot \sum_{\vec{q} \in Q} \vec{q} \in G \right\}\right).$$

Furthermore, let

$$H = \left\{ \frac{bm}{d} \cdot \sum_{\vec{q} \in Q} \vec{q} \;\middle|\; b \in \mathbb{Z} \right\} = G \cap \left\{ p \cdot \sum_{\vec{q} \in Q} \vec{q} \;\middle|\; p \in \mathbb{Q} \right\}$$

and $\pi : G \to G/H$ be the canonical projection. The group G/H is torsion-free, meaning that the identity element is the only element of finite order. Thus, G/H is a finitely generated free abelian group. Let r be the rank of G/H. Hence, there are elements $g_1, g_2, \ldots, g_r \in G$ such that $\pi(\{g_1, g_2, \ldots, g_r\})$ is a basis of G/H. Let $g \in G$. There are numbers $\lambda_1, \lambda_2, \ldots, \lambda_r \in \mathbb{Z}$ with $\pi(g) = \Sigma_{i=1}^{r} \lambda_i \pi(g_i)$. We have that $g - \Sigma_{i=1}^{r} \lambda_i g_i \in H$. So, there is a $b \in \mathbb{Z}$ with

$$g = \frac{bm}{d} \cdot \sum_{\vec{q} \in Q} \vec{q} + \sum_{i=1}^{r} \lambda_i g_i.$$

Thus, the set

$$\left\{ \frac{m}{d} \cdot \sum_{\vec{q} \in Q} \vec{q}, \, g_1, \, g_2, \, \ldots, \, g_r \right\}$$

generates G. Let now $b, \lambda_1, \lambda_2, \ldots, \lambda_r \in \mathbb{Z}$ with

$$\frac{bm}{d} \cdot \sum_{\vec{q} \in Q} \vec{q} + \sum_{i=1}^{r} \lambda_i g_i = \vec{0}.$$

It follows

$$\vec{0} = \pi \left(\frac{bm}{d} \cdot \sum_{\vec{q} \in Q} \vec{q} + \sum_{i=1}^{r} \lambda_i g_i \right) = \sum_{i=1}^{r} \lambda_i \pi(g_i),$$

which implies that for all $i \in \{1, 2, \ldots, r\}$ we have $\lambda_i = 0$. This gives us $b = 0$. Hence,

$$\left\{ \frac{m}{d} \cdot \sum_{\vec{q} \in Q} \vec{q}, \, g_1, \, g_2, \, \ldots, \, g_r \right\}$$

is a basis of the free abelian group G. For every $n \in \mathbb{Z}$ the set

$$B_n = \left\{ \frac{m}{d} \cdot \sum_{\vec{q} \in Q} \vec{q}, \, g_1 + \frac{nm}{d} \cdot \sum_{\vec{q} \in Q} \vec{q}, \, g_2 + \frac{nm}{d} \cdot \sum_{\vec{q} \in Q} \vec{q}, \, \ldots, \, g_r + \frac{nm}{d} \cdot \sum_{\vec{q} \in Q} \vec{q} \right\}$$

is also a basis of G. There is an $n \in \mathbb{N}$ such that $B_n \subseteq \mathbb{N}^k$. We get $L = \mathsf{La}(\vec{c}, B_n)$. $\qquad \square$

Each strongly semirecognizable subset of $\mathbb{N}^k$ that contains the zero vector is a linear set and a lattice:

Lemma 4.27. *Let $k \geq 0$ and $S \subseteq \mathbb{N}^k$ be strongly semirecognizable with $\vec{0} \in S$. Then, it holds*

$$S = \mathsf{L}\left(\vec{0}, \min\left(S \setminus \left\{\vec{0}\right\}\right)\right) = \mathsf{La}\left(\vec{0}, \min\left(S \setminus \left\{\vec{0}\right\}\right)\right).$$

Proof. For $\vec{s}, \vec{t} \in S$ we have $\vec{s} + \vec{t} \in S$. For $\vec{s}, \vec{t} \in S$ with $\vec{t} \leq \vec{s}$ it holds $\vec{s} - \vec{t} \in S$. So, for every finite $F \subseteq S$ we have $\mathsf{L}(\vec{0}, F) \subseteq \mathsf{La}(\vec{0}, F) \subseteq S$. With the definition of minimal elements of subsets of $\mathbb{N}^k$ it follows $S \subseteq \mathsf{L}(\vec{0}, \min(S \setminus \{\vec{0}\}))$. □

Each strongly semirecognizable subset of $\mathbb{N}^k$ equals a unique lattice if we only consider vectors where all components are "large enough:"

Proposition 4.28. *Let $k \geq 0$, $S \subseteq \mathbb{N}^k$ be strongly semirecognizable, and $\vec{s} \in S$. Then, there is a unique lattice L in $\mathbb{N}^k$ with $L_{\geq \vec{s}} = S_{\geq \vec{s}}$. It holds*

$$L = \mathsf{La}\left(\vec{s}, \min\left((S - S) \cap \mathbb{N}^k \setminus \left\{\vec{0}\right\}\right)\right).$$

Proof. Let L be a lattice in $\mathbb{N}^k$ with $L_{\geq \vec{s}} = S_{\geq \vec{s}}$. For all $\vec{x}, \vec{v} \in \mathbb{N}^k$ with $\vec{x} + \vec{v} \in L$ and $\vec{x} + 2\vec{v} \in L$ we have $\vec{x} \in L$. For all $\vec{x} \in L$ there is a finite $P \subseteq \mathbb{N}^k$ with $L = \mathsf{La}(\vec{x}, P)$ and an $n \in \mathbb{N}$ such that $\vec{s} \leq \vec{x} + n \cdot \Sigma_{\vec{p} \in P}\, \vec{p}$. So, we get

$$L = \left\{ \vec{x} \in \mathbb{N}^k \mid \exists \vec{v} \in \mathbb{N}^k : \forall i \in \{1, 2\} : \vec{x} + i\vec{v} \in S_{\geq \vec{s}} \right\}.$$

That proves that there is at most one lattice L in $\mathbb{N}^k$ with $L_{\geq \vec{s}} = S_{\geq \vec{s}}$. By Lemma 4.27 it holds

$$S_{\geq \vec{s}} = \left(\mathsf{La}\left(\vec{s}, \min\left((S - S) \cap \mathbb{N}^k \setminus \left\{\vec{0}\right\}\right)\right)\right)_{\geq \vec{s}}.$$

□

We consider Carathéodory-like decompositions of lattices. By Propositions 4.5 and 4.24 each lattice is a finite union of strongly semirecognizable linear sets. By Theorem 2.22 each lattice is a finite union of linear sets with linearly independent periods. In the following we give a characterization when a lattice is even a finite union of (strongly) semirecognizable linear sets with linearly independent periods. To do so we need some helping results, which are given now.

Lemma 4.29. *Let* $k \geq 0$ *and* V *be a linear subspace of* $\mathbb{R}^k$ *that has a basis which is a subset of* $\mathbb{N}^k$. *Moreover, let* $m \geq 0$ *and the sets* $V_1, V_2, \ldots, V_m \subset V$ *be proper linear subspaces of* V. *Then, it holds* $\mathbb{N}^k \cap V \setminus \cup_{i=1}^m V_i \neq \emptyset$.

Proof. We prove this by induction on m. The statement is clearly true for $m \in \{0, 1\}$. For $m \geq 2$ there is an $\vec{x} \in \mathbb{N}^k \cap V \setminus \cup_{i=1}^{m-1} V_i$ by the induction hypothesis. By basic linear algebra there are $n \geq 0$ and $\vec{y_1}, \vec{y_2}, \ldots, \vec{y_n} \in \mathbb{N}^k$ such that $\{\vec{x}, \vec{y_1}, \vec{y_2}, \ldots, \vec{y_n}\}$ is a basis of V. Because $\cup_{i=1}^{m-1} V_i$ is a topological closed set, there is a $\lambda \in \mathbb{Q}_{>0}$ such that for all $j \in \{1, 2, \ldots, n\}$ we have $\vec{x} + \lambda \vec{y_j} \notin \cup_{i=1}^{m-1} V_i$. Let $\mu \in \mathbb{N}_{>0}$ such that $\mu\lambda \in \mathbb{N}$. Then,

$$B = \{\vec{x}\} \cup \{\, \mu(\vec{x} + \lambda \vec{y_j}) \mid j \in \{1, 2, \ldots, n\}\,\}$$

is a basis of V and it holds $B \cap \cup_{i=1}^{m-1} V_i = \emptyset$. It follows $\emptyset \subset B \setminus V_m \subseteq \mathbb{N}^k \cap V \setminus \cup_{i=1}^m V_i$. $\square$

Using Lemma 4.29 we show the following lemma concerning the linear subspaces spanned by the period sets of lattices and linear sets.

Lemma 4.30. *Let* $k \geq 0$, $\vec{c} \in \mathbb{N}^k$, *and* $P \subseteq \mathbb{N}^k$ *be finite. Furthermore, let* $m > 0$, $\vec{c_1}, \vec{c_2}, \ldots, \vec{c_m} \in \mathbb{N}^k$, *and* $Q_1, Q_2, \ldots, Q_m \subseteq \mathbb{N}^k$ *be finite such that it holds* $\mathsf{La}(\vec{c}, P) \subseteq \cup_{i=1}^m \mathsf{L}(\vec{c_i}, Q_i)$ *and* $\cup_{i=1}^m Q_i \subseteq \mathsf{span}(P)$. *Then, there is an* $i \in \{1, 2, \ldots, m\}$ *fulfilling* $\mathsf{span}(Q_i) = \mathsf{span}(P)$.

Proof. Let $\vec{p} \in \mathsf{span}(P) \cap \mathbb{N}^k$. By basic linear algebra there is a $\lambda \in \mathbb{N}_{>0}$ such that $\lambda\vec{p} \in \mathsf{La}(\vec{0}, P)$. There are $n_1, n_2 \in \mathbb{N}$ with $n_1 < n_2$ and a $j \in \{1, 2, \ldots, m\}$ such that we have $\vec{c} + n_1\lambda\vec{p}, \vec{c} + n_2\lambda\vec{p} \in \mathsf{L}(\vec{c_j}, Q_j)$. Hence, it holds

$$(n_2 - n_1)\lambda\vec{p} = (\vec{c} + n_2\lambda\vec{p}) - (\vec{c} + n_1\lambda\vec{p}) \in \mathsf{La}(0, Q_j),$$

which implies $\vec{p} \in \mathsf{span}(Q_j)$. So, we have shown that $\mathsf{span}(P) \cap \mathbb{N}^k \subseteq \cup_{i=1}^{m} \mathsf{span}(Q_i)$. Now, Lemma 4.29 tells us that there is an $i \in \{1, 2, \ldots, m\}$ with $\mathsf{span}(Q_i) = \mathsf{span}(P)$. □

Now we are ready to prove our result on Carathéodory-like decompositions of lattices. We give a characterization in terms of rational cones when a lattice is a finite union of semirecognizable linear sets with linearly independent periods. Semirecognizable linear sets with linearly independent periods are already strongly semirecognizable by Corollary 4.20.

Theorem 4.31. *For $k \geq 0$, $\vec{c} \in \mathbb{N}^k$, and a finite $P \subseteq \mathbb{N}^k$ the lattice $\mathsf{La}(\vec{c}, P)$ is a finite union of semirecognizable linear sets with linearly independent periods if and only if $\mathsf{span}(P) \cap (\mathbb{R}_{\geq 0})^k$ is a linearly independent rational cone.*

Proof. Assume that $\mathsf{span}(P) \cap (\mathbb{R}_{\geq 0})^k$ is not a linearly independent rational cone and let $m > 0$, $\vec{c_1}, \vec{c_2}, \ldots, \vec{c_m} \in \mathbb{N}^k$, and $Q_1, Q_2, \ldots, Q_m \subseteq \mathbb{N}^k$ be linearly independent subsets of $\mathbb{R}^k$ with $\mathsf{La}(\vec{c}, P) = \cup_{i=1}^{m} \mathsf{L}(\vec{c_i}, Q_i)$. For each $i \in \{1, 2, \ldots, m\}$ we have $\vec{c_i} \in \mathsf{La}(\vec{c}, P)$ which implies $\vec{c_i} - \vec{c} \in \mathsf{span}(P)$. For each $i \in \{1, 2, \ldots, m\}$ and $\vec{p} \in Q_i$ it holds $\vec{p} = ((\vec{c_i} + \vec{p}) - \vec{c}) - (\vec{c_i} - \vec{c}) \in \mathsf{span}(P)$. That gives us $\cup_{i=1}^{m} Q_i \subseteq \mathsf{span}(P)$. By Lemma 4.30 there is an $i \in \{1, 2, \ldots, m\}$ with $\mathsf{span}(Q_i) = \mathsf{span}(P)$. So, $\mathsf{span}(Q_i) \cap (\mathbb{R}_{\geq 0})^k$ is not a linearly independent rational cone. By Proposition 4.16 and Theorem 4.19 we get that $\mathsf{L}(\vec{c_i}, Q_i)$ is not semirecognizable.

Assume now that $\mathsf{span}(P) \cap (\mathbb{R}_{\geq 0})^k$ is a linearly independent rational cone and let $T \subseteq \mathbb{N}^k$ be a linearly independent subset of $\mathbb{R}^k$ with $\mathsf{span}(P) \cap (\mathbb{R}_{\geq 0})^k = \mathsf{cone}(T)$. It follows $\mathsf{span}(P) = \mathsf{span}(T)$. Let $n = |T|$ and $T = \{\vec{t_1}, \vec{t_2}, \ldots, \vec{t_n}\}$. We get

$$T \subseteq \left\{ \sum_{\vec{p} \in P} \lambda_{\vec{p}} \cdot \vec{p} \,\middle|\, \forall \vec{p} \in P : \lambda_{\vec{p}} \in \mathbb{Q} \right\}$$

by basic linear algebra, so for each $i \in \{1, 2, \ldots, n\}$ there are a $\mu_i \in \mathbb{N}_{>0}$ and a $\vec{q_i} \in \mathsf{La}(\vec{0}, P)$ with $\vec{q_i} = \mu_i \vec{t_i}$. Let $Q = \{\vec{q_1}, \vec{q_2}, \ldots, \vec{q_n}\}$. Since $\mathsf{span}(Q) = \mathsf{span}(T)$ and also $\mathsf{cone}(Q) = \mathsf{cone}(T)$, Theorem 4.19 gives us that $\mathsf{L}(\vec{0}, Q)$ is semirecognizable.

Consider the finite set

$$E = \left\{ \sum_{i=1}^{n} \nu_i \vec{q}_i \;\middle|\; \forall i \in \{1,2,\ldots,n\} : (\nu_i \in \mathbb{R} \wedge 0 \leq \nu_i < 1) \right\} \cap \mathsf{La}\left(\vec{0}, P\right).$$

Let $M = \min\left(\mathsf{La}(\vec{c}, P)\right)$ and

$$S = \bigcup_{\vec{m} \in M, \vec{e} \in E} \mathsf{L}(\vec{m} + \vec{e}, Q).$$

Because of $M \subseteq \mathsf{La}(\vec{c}, P)$ and $E, Q \subseteq \mathsf{La}(\vec{0}, P)$, we have $S \subseteq \mathsf{La}(\vec{c}, P)$. In the following we show $\mathsf{La}(\vec{c}, P) = S$.

Let $\vec{x} \in \mathsf{La}(\vec{c}, P)$. There is an $\vec{m} \in M$ with $\vec{m} \leq \vec{x}$. Because of $\vec{x} - \vec{m} \in \mathsf{La}(\vec{0}, P) \subseteq \mathsf{cone}(Q)$ there are $\lambda_1, \lambda_2, \ldots, \lambda_n \in \mathbb{R}_{\geq 0}$ with $\vec{x} - \vec{m} = \Sigma_{i=1}^{n} \lambda_i \vec{q}_i$. We have

$$\sum_{i=1}^{n} \lfloor \lambda_i \rfloor \vec{q}_i \in \mathsf{L}\left(\vec{0}, Q\right) \subseteq \mathsf{La}\left(\vec{0}, P\right)$$

and

$$\vec{x} - \vec{m} - \sum_{i=1}^{n} \lfloor \lambda_i \rfloor \vec{q}_i = \sum_{i=1}^{n} (\lambda_i - \lfloor \lambda_i \rfloor) \vec{q}_i \in E.$$

It follows $\vec{x} \in S$. So we have shown $\mathsf{La}(\vec{c}, P) = S$, which proves the theorem. $\square$

We can generalize Theorem 4.31 from lattices to strongly semirecognizable subsets of $\mathbb{N}^k$:

Corollary 4.32. *Let $k \geq 0$ and $S \subseteq \mathbb{N}^k$ be strongly semirecognizable. Then, S is a finite union of semirecognizable linear sets with linearly independent periods if and only if*

$$\mathsf{span}\left((S - S) \cap \mathbb{N}^k\right) \cap \left(\mathbb{R}_{\geq 0}\right)^k$$

is a linearly independent rational cone.

Proof. By Proposition 4.28 for every $\vec{s} \in S$ the set $S_{\geq \vec{s}}$ equals

$$\left(\mathsf{La}\left(\vec{s}, \min\left((S - S) \cap \mathbb{N}^k \setminus \left\{\vec{0}\right\}\right)\right)\right)_{\geq \vec{s}}$$
$$= \{\vec{s}\} + \mathsf{La}\left(\vec{0}, \min\left((S - S) \cap \mathbb{N}^k \setminus \left\{\vec{0}\right\}\right)\right).$$

It follows

$$\mathsf{span}\left((S-S)\cap\mathbb{N}^k\right)=\mathsf{span}\left(\min\left((S-S)\cap\mathbb{N}^k\setminus\left\{\vec{0}\right\}\right)\right)$$

and

$$S=\bigcup_{\vec{s}\in\min(S)}S_{\geq\vec{s}}=\min(S)+\mathsf{La}\left(\vec{0},\min\left((S-S)\cap\mathbb{N}^k\setminus\left\{\vec{0}\right\}\right)\right).$$

If

$$\mathsf{span}\left((S-S)\cap\mathbb{N}^k\right)\cap(\mathbb{R}_{\geq0})^k$$

is a linearly independent rational cone, Theorem 4.31 tells us that S is a finite union of semirecognizable linear sets with linearly independent periods.

Let $n\geq0$ and $\vec{c},\vec{x},\vec{p_1},\vec{p_2},\ldots,\vec{p_n}\in\mathbb{N}^k$. Then, $(\mathsf{L}(\vec{c},\{\vec{p_1},\vec{p_2},\ldots,\vec{p_n}\}))_{\geq\vec{x}}$ equals

$$\mathsf{L}\left(\left\{\vec{c}+\sum_{i=1}^n\lambda_i\vec{p_i}\;\middle|\;\vec{\lambda}\in\min\left(\left\{\vec{\lambda}\in\mathbb{N}^n\;\middle|\;\vec{x}\leq\vec{c}+\sum_{i=1}^n\lambda_i\vec{p_i}\right\}\right)\right\},\right.$$

$$\left.\{\vec{p_1},\vec{p_2},\ldots,\vec{p_n}\}\right).$$

Thus, if S is a finite union of semirecognizable linear sets with linearly independent periods,

$$\mathsf{La}\left(\vec{0},\min\left((S-S)\cap\mathbb{N}^k\setminus\left\{\vec{0}\right\}\right)\right)$$

is as well a finite union of semirecognizable linear sets with linearly independent periods and by Theorem 4.31 the set

$$\mathsf{span}\left((S-S)\cap\mathbb{N}^k\right)\cap(\mathbb{R}_{\geq0})^k$$

is a linearly independent rational cone. $\qquad\square$

The proofs of Theorem 4.31 and Corollary 4.32 reveal:

Corollary 4.33. *Let $k\geq0$ and $S\subseteq\mathbb{N}^k$ be a strongly semirecognizable set that is a finite union of semirecognizable linear sets with linearly independent periods. Then, there are a finite $C\subseteq\mathbb{N}^k$ and a $P\subseteq\mathbb{N}^k$ which is a linearly independent subset of $\mathbb{R}^k$ such that $S=\mathsf{L}(C,P)$ and $\mathsf{L}(\vec{0},P)$ is strongly semirecognizable.* $\qquad\square$

For each subset of a monoid all syntactic equivalence classes are strongly semirecognizable subsets of the monoid. A subset of a monoid is semirecognizable if and only if there are only finitely many syntactic equivalence classes for which the elements are contained in the subset. A semirecognizable subset S of $\mathbb{N}^k$ is a finite union of semirecognizable linear sets with linearly independent periods if and only if all syntactic equivalence classes for which the elements are contained in S are finite unions of semirecognizable linear sets with linearly independent periods:

Proposition 4.34. *Let $k \geq 0$ and $S \subseteq \mathbb{N}^k$ be semirecognizable. Then, S is a finite union of semirecognizable linear sets with linearly independent periods if and only if each element of $S/\sim_S$ is a finite union of semirecognizable linear sets with linearly independent periods.*

Proof. Let $m > 0$, $\vec{c_1}, \vec{c_2}, \ldots, \vec{c_m} \in \mathbb{N}^k$, and $P_1, P_2, \ldots, P_m \subseteq \mathbb{N}^k$ be linearly independent subsets of $\mathbb{R}^k$ such that $S = \cup_{i=1}^m \mathsf{L}(\vec{c_i}, P_i)$ and for each $i \in \{1, 2, \ldots, m\}$ the set $\mathsf{L}(\vec{c_i}, P_i)$ is semirecognizable. For each $i \in \{1, 2, \ldots, m\}$ let $P_i = \{\vec{p_{i,1}}, \vec{p_{i,2}}, \ldots, \vec{p_{i,|P_i|}}\}$. Furthermore let $T \in S/\sim_S$ and

$$X = \left\{ j \in \{1, 2, \ldots, k\} \,\middle|\, \forall \vec{v} \in (T - T) \cap \mathbb{N}^k : v_j = 0 \right\}.$$

For each $i \in \{1, 2, \ldots, m\}$ let $Q_i = \{ \vec{p} \in P_i \mid \forall j \in X : p_j = 0 \}$,

$$M_i = \min \left(\left\{ \vec{\lambda} \in \mathbb{N}^{|P_i|} \,\middle|\, \vec{c_i} + \sum_{j=1}^{|P_i|} \lambda_j \vec{p_{i,j}} \in T \right\} \right),$$

and $N_i = \{ \vec{c_i} + \Sigma_{j=1}^{|P_i|} \lambda_j \vec{p_{i,j}} \mid \vec{\lambda} \in M_i \}$. So, for all $i \in \{1, 2, \ldots, m\}$ we have $T \cap \mathsf{L}(\vec{c_i}, P_i) \subseteq \mathsf{L}(N_i, Q_i)$.

Set

$$Y = \left\{ i \in \{1, 2, \ldots, m\} \mid T \cap \mathsf{L}(\vec{c_i}, P_i) \neq \emptyset \right\}.$$

Let $i \in Y$ and $\vec{q} \in Q_i$. There are $r, s \in \mathbb{N}$ with $r < s$ and a $U \in S/\sim_S$ such that it holds $\vec{c_i} + r\vec{q} \in U$ and $\vec{c_i} + s\vec{q} \in U$. Now, there is a $\vec{v} \in T$ with $\vec{c_i} + r\vec{q} \leq \vec{v}$. We get $\vec{v} + (s - r)\vec{q} \in T$ because $(\vec{v} + (s - r)\vec{q}) - (\vec{c_i} + s\vec{q}) = \vec{v} - (\vec{c_i} + r\vec{q})$. It follows $\vec{q} \in \mathsf{span}\,((T - T) \cap \mathbb{N}^k)$. Thus, we have shown $\cup_{i \in Y} Q_i \subseteq \mathsf{span}\,((T - T) \cap \mathbb{N}^k)$. By Proposition 4.28 and Lemma 4.30 there is an $i \in Y$ with $\mathsf{span}(Q_i) = \mathsf{span}\,((T - T) \cap \mathbb{N}^k)$. By

Corollary 4.23 and Theorem 4.19 we have $\mathsf{cone}(Q_i) = \mathsf{span}(Q_i) \cap (\mathbb{R}_{\geq 0})^k$. Corollary 4.32 implies that T is a finite union of semirecognizable linear sets with linearly independent periods. □

Because of Theorem 4.31 and Corollary 4.32 we investigate when the vectors with only non-negative components in the linear subspace of $\mathbb{R}^k$ spanned by a subset of $\mathbb{N}^k$ form a linearly independent rational cone. Intuitively one might think that this is always the case, but it turns out that that is only true in dimension $k \leq 3$:

Theorem 4.35. *Let $k \geq 0$ and $n \in \{0, 1, \ldots, k\}$. Then, the condition that for each $S \subseteq \mathbb{N}^k$ with $\dim(\mathsf{span}(S)) = n$ the set $\mathsf{span}(S) \cap (\mathbb{R}_{\geq 0})^k$ is a linearly independent rational cone holds if and only if $n \in \{0, 1, 2, k\}$.*

Proof. For $S \subseteq \mathbb{N}^k$ with $\dim(\mathsf{span}(S)) = 0$ we have

$$\mathsf{span}(S) \cap (\mathbb{R}_{\geq 0})^k = \left\{ \vec{0} \right\} = \mathsf{cone}\left(\emptyset \right).$$

For $S \subseteq \mathbb{N}^k$ with $\dim(\mathsf{span}(S)) = 1$ and a $\vec{v} \in S \setminus \{\vec{0}\}$ it holds $\mathsf{span}(S) \cap (\mathbb{R}_{\geq 0})^k = \mathsf{cone}(\{v\})$. For $S \subseteq \mathbb{N}^k$ with $\dim(\mathsf{span}(S)) = k$ we get

$$\mathsf{span}(S) \cap (\mathbb{R}_{\geq 0})^k = \mathsf{cone}\left(\{\vec{e_{k,1}}, \vec{e_{k,2}}, \ldots, \vec{e_{k,k}}\} \right).$$

Let now $S \subseteq \mathbb{N}^k$ with $\dim(\mathsf{span}(S)) = 2 < k$ and let $\{\vec{v}, \vec{w}\} \subseteq S$ be a basis of $\mathsf{span}(S)$. Let

$$\lambda = \min\left(\left\{ \frac{v_j}{w_j} \ \middle|\ j \in \{1, 2, \ldots, k\},\ w_j > 0 \right\} \right)$$

and set $\vec{v'} := \vec{v} - \lambda \vec{w}$. So $\{\vec{v'}, \vec{w}\}$ is a basis of $\mathsf{span}(S)$. For each $j \in \{1, 2, \ldots, k\}$ with $\vec{w}_j > 0$ it holds

$$v'_j = v_j - \lambda w_j \geq v_j - \frac{v_j}{w_j} \cdot w_j = 0.$$

This gives us $\vec{v'} \in (\mathbb{Q}_{\geq 0})^k$. Now let

$$\mu = \min\left(\left\{ \frac{w_j}{v'_j} \ \middle|\ j \in \{1, 2, \ldots, k\},\ v'_j > 0 \right\} \right)$$

and set $\vec{w'} := \vec{w} - \mu\vec{v'}$. Thus $\{\vec{v'}, \vec{w'}\}$ is a basis of $\mathsf{span}(S)$. For each $j \in \{1, 2, \ldots, k\}$ with $\vec{v'}_j > 0$ we have

$$w'_j = w_j - \mu v'_j \geq w_j - \frac{w_j}{v'_j} \cdot v'_j = 0.$$

That implies $\vec{w'} \in (\mathbb{Q}_{\geq 0})^k$.

For all $j \in \{1, 2, \ldots, k\}$ with $w_j > 0$ and $\frac{v_j}{w_j} = \lambda$ it holds $v'_j = 0$ and also $w'_j > 0$. For all $j \in \{1, 2, \ldots, k\}$ with $v'_j > 0$ and $\frac{w_j}{v'_j} = \mu$ we have $w'_j = 0$. Let $\nu, \xi \in \mathbb{N}_{>0}$ with $\nu\vec{v'}, \xi\vec{w'} \in \mathbb{N}^k$. We get

$$\mathsf{span}(S) \cap (\mathbb{R}_{\geq 0})^k = \mathsf{span}\left(\left\{\nu\vec{v'}, \xi\vec{w'}\right\}\right) \cap (\mathbb{R}_{\geq 0})^k = \mathsf{cone}\left(\left\{\nu\vec{v'}, \xi\vec{w'}\right\}\right).$$

Let now $n \in \{3, 4, \ldots, k-1\}$ and consider the 4×4 matrix

$$A = \begin{pmatrix} 0 & 1 & 1 & 2 \\ 1 & 0 & 2 & 1 \\ 1 & 2 & 0 & 1 \\ 2 & 1 & 1 & 0 \end{pmatrix}.$$

For all $p, q \geq 0$ let I_p be the $p \times p$ identity matrix, $J_{p,q}$ be the $p \times q$ all-ones matrix, and $0_{p,q}$ be the $p \times q$ zero matrix. Let $S_{k,n}$ be the $k \times (n+1)$ block matrix

$$S_{k,n} = \begin{pmatrix} A & J_{4,n-3} \\ J_{n-3,4} & J_{n-3,n-3} - I_{n-3} \\ 0_{k-n-1,4} & 0_{k-n-1,n-3} \end{pmatrix}.$$

We refer to the columns of $S_{k,n}$ as $\vec{v_1}, \vec{v_2}, \ldots, \overrightarrow{v_{n+1}} \in \mathbb{N}^k$. It holds $\vec{v_1} = \vec{v_2} + \vec{v_3} - \vec{v_4}$. Let $\lambda_2, \lambda_3, \ldots, \lambda_{n+1} \in \mathbb{R}$, set $\vec{w} = \sum_{i=2}^{n+1} \lambda_i \vec{v_i}$, and assume that $\vec{w} = \vec{0}$. Then we have $0 = w_1 - w_4 = 2\lambda_4$. For all $i \in \{5, 6, \ldots, n+1\}$ we have $0 = w_4 - w_i = \lambda_i$. Now it is easy to see that we also have $\lambda_2 = \lambda_3 = 0$. So the vectors $\vec{v_2}, \vec{v_3}, \ldots, \overrightarrow{v_{n+1}}$ are linearly independent and the matrix $S_{k,n}$ has rank n.

Let $V_{k,n}$ be the set of columns of $S_{k,n}$, assume that $\mathsf{span}\left(V_{k,n}\right) \cap (\mathbb{R}_{\geq 0})^k$ is a linearly independent rational cone, and let $B = \{\vec{b_1}, \vec{b_2}, \ldots, \vec{b_n}\} \subset \mathbb{N}^k$ be linearly independent with

$$\mathsf{span}\left(V_{k,n}\right) \cap (\mathbb{R}_{\geq 0})^k = \mathsf{cone}(B).$$

Furthermore, let $(b_{j,i})_{j \in \{1,2,\ldots,k\}, i \in \{1,2,\ldots,n\}}$ be the $k \times n$ matrix whose columns are from left to right $\vec{b_1}, \vec{b_2}, \ldots, \vec{b_n}$. For all $j \in \{n+2, n+3, \ldots, k\}$ and $i \in \{1, 2, \ldots, n\}$ we have $b_{j,i} = 0$.

Assume that there is an $i_0 \in \{1, 2, \ldots, n\}$ such that for all $j \in \{1, 2, \ldots, n+1\}$ there is an $i \in \{1, 2, \ldots, n\} \setminus \{i_0\}$ with $b_{j,i} > 0$. Set $\vec{x} = \Sigma_{i \in \{1,2,\ldots,n\} \setminus \{i_0\}} \vec{b_i}$. For all $j \in \{1, 2, \ldots, n+1\}$ we have $x_j > 0$. Let

$$\lambda = \min \left\{ \frac{x_j}{b_{j,i_0}} \;\middle|\; j \in \{1, 2, \ldots, k\}, \; b_{j,i_0} > 0 \right\} > 0$$

and set $\vec{x'} = \vec{x} - \lambda \vec{b_{i_0}}$. For each $j \in \{1, 2, \ldots, k\}$ with $b_{j,i_0} > 0$ we get

$$x'_j = x_j - \lambda b_{j,i_0} \geq x_j - \frac{x_j}{b_{j,i_0}} \cdot b_{j,i_0} = 0.$$

This is a contradiction to

$$\mathsf{span}(B) \cap (\mathbb{R}_{\geq 0})^k = \mathsf{span}\,(V_{k,n}) \cap (\mathbb{R}_{\geq 0})^k = \mathsf{cone}(B)$$

because $\vec{x'} = \Sigma_{i \in \{1,2,\ldots,n\} \setminus \{i_0\}} \vec{b_i} - \lambda \vec{b_{i_0}}$. Hence, for all $i \in \{1, 2, \ldots, n\}$ there is a $j_i \in \{1, 2, \ldots, n+1\}$ such that for all $i' \in \{1, 2, \ldots, n\} \setminus \{i\}$ we have $b_{j_i, i'} = 0$. It follows that for all $i \in \{1, 2, \ldots, n\}$ it holds $b_{j_i, i} > 0$. Let

$$C = \{\, j_i \mid i \in \{1, 2, \ldots, n\} \,\} \subset \{1, 2, \ldots, n+1\}.$$

We get $|C| = n$. Let m be the only element of $\{1, 2, \ldots, n+1\} \setminus C$. From what we have shown about the matrix $(b_{j,i})_{j \in \{1,2,\ldots,k\}, i \in \{1,2,\ldots,n\}}$ it follows that for every $\vec{y} \in \mathsf{span}(B)$ it holds: if for all $j \in C$ we have $y_j \geq 0$, then we also have $y_m \geq 0$.

Let now $i \in C$ and set

$$\vec{y} = \vec{v_m} - \frac{1}{3} \cdot \vec{v_i} \in \mathsf{span}(V_{k,n}) = \mathsf{span}(B).$$

This gives us $y_m < 0$. For all $j \in C$ we have $y_j \geq 1 - \frac{1}{3} \cdot 2 = \frac{1}{3}$, a contradiction. So, $\mathsf{span}\,(V_{k,n}) \cap (\mathbb{R}_{\geq 0})^k$ is not a linearly independent rational cone. $\qquad \square$

4.3 Finite Unions of Semirecognizable Sets

We characterize when an arbitrary subset of $\mathbb{N}^k$ is a finite union of semirecognizable subsets of $\mathbb{N}^k$. For our characterization we need the following notion. For $k \geq 0$ an $S \subseteq \mathbb{N}^k$ is called a *quasi-lattice* if there are an $\vec{y} \in \mathbb{N}^k$ and a finite set X of lattices in $\mathbb{N}^k$ such that $S_{\geq \vec{y}} = (\cup X)_{\geq \vec{y}}$. We can identify a pattern of two linear sets formed by three vectors that gives a sufficient condition for the property of a subset of $\mathbb{N}^k$ to not be a quasi-lattice:

Proposition 4.36. *Let $k \geq 0$, $S \subseteq \mathbb{N}^k$, $\vec{u}, \vec{w} \in \mathbb{N}^k$, and $\vec{v} \in (\mathbb{N}_{>0})^k$ with $\mathsf{L}(\vec{u}, \{\vec{v}\}) \cap S = \emptyset$ and $\mathsf{L}(\vec{u} + \vec{w}, \{\vec{v}, \vec{w}\}) \subseteq S$. Then, S is not a quasi-lattice.*

Proof. Let $\vec{y} \in \mathbb{N}^k$. There is a $\lambda \in \mathbb{N}$ with $\vec{u} + \lambda \vec{v} \geq \vec{y}$. Consider a number $m > 0$, $\vec{c_1}, \vec{c_2}, \ldots, \vec{c_m} \in \mathbb{N}^k$, and finite $P_1, P_2, \ldots, P_m \subseteq \mathbb{N}^k$ such that

$$\mathsf{L}(\vec{u} + \vec{w} + \lambda \vec{v}, \{\vec{w}\}) \subseteq \bigcup_{j=1}^{m} \mathsf{La}(\vec{c_j}, P_j).$$

W.l.o.g. there is an $n > 0$ with $n \leq m$ such that for all $j \in \{1, 2, \ldots, m\}$ the inequality

$$|\mathsf{L}(\vec{u} + \vec{w} + \lambda \vec{v}, \{\vec{w}\}) \cap \mathsf{La}(\vec{c_j}, P_j)| \geq 2$$

is equivalent to $j \leq n$. For all $j \in \{1, 2, \ldots, n\}$ let M_j be the minimum of

$$\{\, \mu_2 - \mu_1 \mid \mu_1, \mu_2 \in \mathbb{N}, 0 < \mu_1 < \mu_2, \forall i \in \{1, 2\} : \vec{u} + \lambda \vec{v} + \mu_i \vec{w} \in \mathsf{La}(\vec{c_j}, P_j) \,\}$$

and let M be the lowest common multiple of the M_j. For all $j \in \{1, 2, \ldots, n\}$ and $\nu \geq 0$ we have $\vec{u} + \lambda \vec{v} \in \mathsf{La}(\vec{c_j}, P_j)$ if and only if $\vec{u} + \lambda \vec{v} + \nu M \vec{w} \in \mathsf{La}(\vec{c_j}, P_j)$. It follows $\vec{u} + \lambda \vec{v} \in \cup_{j=1}^{n} \mathsf{La}(\vec{c_j}, P_j)$. Since

$$\mathsf{L}(\vec{u} + \vec{w} + \lambda \vec{v}, \{\vec{w}\}) \subseteq S,$$

but $\vec{u} + \lambda \vec{v} \notin S$, S is not a quasi-lattice. $\qquad \square$

Example 4.37. Consider the set $S = \{\, (n, m) \in \mathbb{N}^2 \mid n \neq m \,\}$. With $\vec{u} = (0, 0)$, $v = (1, 1)$, and $w = (1, 0)$ we have $\mathsf{L}(\vec{u}, \{\vec{v}\}) = \mathbb{N}^2 \setminus S$ and

$$\mathsf{L}(\vec{u} + \vec{w}, \{\vec{v}, \vec{w}\}) = \{\, (n, m) \in \mathbb{N}^2 \mid n > m \,\} \subset S.$$

Hence, S is not a quasi-lattice by Proposition 4.36.

We call subsets of $\mathbb{N}^k$ that allow a pattern as in Proposition 4.36 anti-lattices. If they even allow a pattern that begins at the origin, we call them 0-anti-lattices. Let $k \geq 0$ and $S \subseteq \mathbb{N}^k$. If there are vectors $\vec{u}, \vec{w} \in \mathbb{N}^k$ and $\vec{v} \in (\mathbb{N}_{>0})^k$ with $\mathsf{L}(\vec{u}, \{\vec{v}\}) \cap S = \emptyset$ and $\mathsf{L}(\vec{u} + \vec{w}, \{\vec{v}, \vec{w}\}) \subseteq S$, the set S is called an *anti-lattice*. If there are $\vec{w} \in \mathbb{N}^k$ and $\vec{v} \in (\mathbb{N}_{>0})^k$ with $\mathsf{L}(\vec{0}, \{\vec{v}\}) \cap S = \emptyset$ and $\mathsf{L}(\vec{w}, \{\vec{v}, \vec{w}\}) \subseteq S$, the set S is called a *0-anti-lattice*. Thus, the set $\{(n, m) \in \mathbb{N}^2 \mid n \neq m\}$ from Example 4.37 is a 0-anti-lattice. The following property of anti-lattices is used later on.

Lemma 4.38. *Let $k \geq 0$ and $S \subseteq \mathbb{N}^k$. Then, S is an anti-lattice if and only if there is an $\vec{x} \in \mathbb{N}^k$ such that $S + \{\vec{x}\}$ is a 0-anti-lattice.*

Proof. The lemma is clearly true for $k = 0$, so let $k > 0$. Assume that there is a vector $\vec{x} \in \mathbb{N}^k$ such that $S + \{\vec{x}\}$ is a 0-anti-lattice. Thus, there are $\vec{w} \in \mathbb{N}^k$ and $\vec{v} \in (\mathbb{N}_{>0})^k$ with $\mathsf{L}(\vec{0}, \{\vec{v}\}) \cap (S + \{\vec{x}\}) = \emptyset$ and $\mathsf{L}(\vec{w}, \{\vec{v}, \vec{w}\}) \subseteq S + \{\vec{x}\}$. Let $\lambda \in \mathbb{N}$ with $\lambda \vec{v} \geq \vec{x}$. We get $\mathsf{L}(\lambda \vec{v} - \vec{x}, \{\vec{v}\}) \cap S = \emptyset$ and $\mathsf{L}(\lambda \vec{v} - \vec{x} + \vec{w}, \{\vec{v}, \vec{w}\}) \subseteq S$. So, S is an anti-lattice.

Assume now that S is an anti-lattice. There are $\vec{u}, \vec{w} \in \mathbb{N}^k$, and $\vec{v} \in (\mathbb{N}_{>0})^k$ such that $\mathsf{L}(\vec{u}, \{\vec{v}\}) \cap S = \emptyset$ and $\mathsf{L}(\vec{u} + \vec{w}, \{\vec{v}, \vec{w}\}) \subseteq S$. Let $\vec{x} \in \mathbb{N}^k \setminus \{\vec{0}\}$ and $\lambda \in \mathbb{N}$ with $\vec{u} + \vec{x} = \lambda \vec{v}$. We get $\mathsf{L}(\vec{0}, \{\lambda \vec{v}\}) \cap (S + \{\vec{x}\}) = \emptyset$ and

$$\mathsf{L}(\lambda \vec{v} + \vec{w}, \{\lambda \vec{v}, \lambda \vec{v} + \vec{w}\}) \subseteq S + \{\vec{x}\}.$$

So, $S + \{\vec{x}\}$ is a 0-anti-lattice. $\square$

For semilinear sets the implication from Proposition 4.36 becomes an equivalence:

Proposition 4.39. *Let $k \geq 0$ and $S \subseteq \mathbb{N}^k$ be a semilinear set. Then, S is a quasi-lattice if and only if S is not an anti-lattice.*

Proof. Assume that S is not a quasi-lattice. As a semilinear set S is a finite union of linear sets, so there are a $\vec{c} \in \mathbb{N}^k$ and a finite $P \subseteq \mathbb{N}^k$ such that $\mathsf{L}(\vec{c}, P) \subseteq S$ and for all $\vec{y} \in \mathbb{N}^k$ it holds $(\mathsf{La}(\vec{c}, P))_{\geq \vec{y}} \not\subseteq S$. Let $P = \{\vec{p_1}, \vec{p_2}, \ldots, \vec{p}_{|P|}\}$ and consider an arbitrary $\vec{y} \in \mathbb{N}^k$. Then, let $N_{\vec{y}}$

be the non-empty set

$$\left\{ \vec{\lambda} \in \mathbb{N}^{|P|} \ \middle|\ \exists \vec{\mu} \in \mathbb{N}^{|P|} : \sum_{j=1}^{|P|} (\mu_j - \lambda_j) \vec{p_j} \in (\mathbb{N}^k \setminus S)_{\geq \vec{y}} \right\}$$

and $M_{\vec{y}} = \min(N_{\vec{y}})$. For all $(\lambda_1, \lambda_2, \ldots, \lambda_{|P|}) \in M_{\vec{y}}$ there are a vector $(\mu_1, \mu_2, \ldots, \mu_{|P|}) \in \mathbb{N}^{|P|}$ and $j_0 \in \{1, 2, \ldots, |P|\}$ such that on the one hand $\Sigma_{j=1}^{|P|}(\mu_j - \lambda_j)\vec{p_j} \in (\mathbb{N}^k \setminus S)_{\geq \vec{y}}$ and on the other hand for all $\alpha > 0$ it holds $\alpha \vec{p_{j_0}} + \Sigma_{j=1}^{|P|}(\mu_j - \lambda_j)\vec{p_j} \in S$.

For $m \geq 0$ let $\vec{y_m} = (m, m, \ldots, m) \in \mathbb{N}^k$. Let now j_0 be in the non-empty set

$$\bigcap_{m=0}^{\infty} \left\{ j \in \{1, 2, \ldots, |P|\} \ \middle|\ \exists \vec{x} \in (\mathbb{N}^k \setminus S)_{\geq \vec{y_m}} : \forall \alpha > 0 : \vec{x} + \alpha \vec{p_j} \in S \right\}.$$

The set

$$T = \left\{ \vec{x} \in \mathbb{N}^k \setminus S \ \middle|\ \forall \alpha > 0 : \vec{x} + \alpha \vec{p_{j_0}} \in S \right\}$$

is semilinear by Theorem 2.26. So, there are $\ell \geq 0$, $\vec{d_1}, \vec{d_2}, \ldots, \vec{d_\ell} \in \mathbb{N}^k$, and finite sets $Q_1, Q_2, \ldots, Q_\ell \subseteq \mathbb{N}^k$ such that $T = \cup_{h=1}^{\ell} \mathsf{L}(\vec{d_h}, Q_h)$. Since for each $\vec{y} \in \mathbb{N}^k$ there is an $\vec{x} \in T$ with $\vec{x} \geq \vec{y}$, there exists an $h_0 \in \{1, 2, \ldots, \ell\}$ such that for all $j \in \{1, 2, \ldots, k\}$ it holds $\pi_{k,\{j\}}\left(\Sigma_{\vec{q} \in Q_{h_0}} \vec{q}\right) > 0$. We have

$$\mathsf{L}\left(\vec{d_{h_0}}, \left\{\sum_{\vec{q} \in Q_{h_0}} \vec{q}\right\}\right) \cap S = \emptyset$$

and

$$\mathsf{L}\left(\vec{d_{h_0}} + \vec{p_{j_0}}, \left\{\vec{p_{j_0}}, \sum_{\vec{q} \in Q_{h_0}} \vec{q}\right\}\right) \subseteq S.$$

Thus, S is an anti-lattice. Our proposition follows with Proposition 4.36. $\square$

Now we give a characterization in terms of quasi-lattices when an arbitrary subset of $\mathbb{N}^k$ is a finite union of semirecognizable subsets of $\mathbb{N}^k$. A finite union of semirecognizable subsets of $\mathbb{N}^k$ is already a finite union of strongly semirecognizable linear subsets of $\mathbb{N}^k$ by Proposition 4.5. In dimension $k \leq 3$ each finite union of semirecognizable subsets of $\mathbb{N}^k$ is even a finite union of strongly semirecognizable linear sets with linearly independent periods by Corollaries 4.20 and 4.32 and Theorem 4.35.

Proposition 4.40. *Let $k \geq 0$ and $S \subseteq \mathbb{N}^k$. Then, S is a finite union of semirecognizable subsets of $\mathbb{N}^k$ if and only if for all subsets $T \subseteq \{1, 2, \ldots, k\}$ and vectors $\vec{x} \in \mathbb{N}^{|T|}$ the set $\pi_{k,\{1,2,\ldots,k\}\setminus T}\left(S \cap \pi_{k,T}^{-1}(\{\vec{x}\})\right)$ is a quasi-lattice.*

Proof. If S is a finite union of semirecognizable subsets of $\mathbb{N}^k$, then for all $T \subseteq \{1, 2, \ldots, k\}$ and $\vec{x} \in \mathbb{N}^{|T|}$ the set $\pi_{k,\{1,2,\ldots,k\}\setminus T}\left(S \cap \pi_{k,T}^{-1}(\{\vec{x}\})\right)$ is clearly a finite union of semirecognizable subsets of $\mathbb{N}^{k-|T|}$. For all $k' \geq 0$ each finite union of semirecognizable subsets of $\mathbb{N}^{k'}$ is a quasi-lattice by Proposition 4.28.

We now prove by induction on k that if for all $T \subseteq \{1, 2, \ldots, k\}$ and $\vec{x} \in \mathbb{N}^{|T|}$ the set $\pi_{k,\{1,2,\ldots,k\}\setminus T}\left(S \cap \pi_{k,T}^{-1}(\{\vec{x}\})\right)$ is a quasi-lattice, then S is a finite union of semirecognizable subsets of $\mathbb{N}^k$. This is true for $k = 0$, so let $k > 0$ and assume that for all $T \subseteq \{1, 2, \ldots, k\}$ and $\vec{x} \in \mathbb{N}^{|T|}$ the set $\pi_{k,\{1,2,\ldots,k\}\setminus T}\left(S \cap \pi_{k,T}^{-1}(\{\vec{x}\})\right)$ is a quasi-lattice. Choosing $T = \emptyset$ gives us that S is a quasi-lattice. So, there exist $\vec{y} \in \mathbb{N}^k$ and a finite set X of lattices in $\mathbb{N}^k$ such that $S_{\geq \vec{y}} = (\cup X)_{\geq \vec{y}}$. By Proposition 4.24 the set $S_{\geq \vec{y}}$ is a finite union of semirecognizable subsets of $\mathbb{N}^k$. It holds

$$S \setminus S_{\geq \vec{y}} = \left\{\, \vec{z} \in S \mid \vec{y} \not\leq \vec{z} \,\right\} = \bigcup_{p=1}^{k} \bigcup_{q=0}^{y_p-1} \left\{\, \vec{z} \in S \mid z_p = q \,\right\}.$$

Now let $p \in \{1, 2, \ldots, k\}$ and $q \in \{0, 1, \ldots, y_p-1\}$ and define the map $\iota_{p,q} : \mathbb{N}^{k-1} \to \mathbb{N}^k$ as

$$\iota_{p,q}(\vec{x}) = (x_1, x_2, \ldots, x_{p-1}, q, x_p, x_{p+1}, \ldots, x_{k-1}).$$

By the induction hypothesis $\iota_{p,q}^{-1}(\{\, \vec{z} \in S \mid z_p = q \,\})$ is a finite union of semirecognizable subsets of $\mathbb{N}^{k-1}$. That implies that the set $\{\, \vec{z} \in S \mid z_p = q \,\}$ is a finite union of semirecognizable subsets of $\mathbb{N}^k$. Thus, S is a finite union of semirecognizable subsets of $\mathbb{N}^k$. $\square$

By Propositions 2.30, 4.5, and 4.28, a subset of $\mathbb{N}$ is a quasi-lattice if and only if it is semilinear. So, in dimension $k \leq 2$ the characterization from Proposition 4.40 becomes simpler:

Proposition 4.41. *Let $k \in \{0, 1, 2\}$ and $S \subseteq \mathbb{N}^k$. Then, S is a finite union of semirecognizable subsets of $\mathbb{N}^k$ if and only if S is semilinear and a quasi-lattice.* $\qquad\square$

4.4 Parikh-Preimages of Semirecognizable Sets

We study properties of Parikh-preimages of semirecognizable subsets of $\mathbb{N}^k$. This family of languages coincides with the permutation closed semirecognizable languages, which are a proper subfamily of the permutation closed semilinear languages by Proposition 4.5 and Example 4.1. In the paper *Characterization and Complexity Results on Jumping Finite Automata* by Fernau *et al.* it is argued at the beginning of Section 8 that each permutation closed semilinear language is in $\mathbf{NST}(\log(n), n)$ [29]. Using Theorem 2.22, we can improve this:

Corollary 4.42. *Each permutation closed semilinear language is contained in $\mathbf{DST}(\log(n), n)$.*

Proof. Let Σ be an alphabet, $k = |\Sigma|$, and $L \subseteq \Sigma^*$ be permutation closed and semilinear. By Theorem 2.22 there are a finite set I and for each $i \in I$ a vector $\vec{c}_i \in \mathbb{N}^k$ and a subset $P_i \subseteq \mathbb{N}^k$ which is linearly independent as a subset of the k-dimensional Euclidean space such that $\psi_\Sigma(L) = \bigcup_{i \in I} \mathsf{L}(\vec{c}_i, P_i)$. For each $i \in I$ let $P_i = \{\overrightarrow{p_{i,1}}, \overrightarrow{p_{i,2}}, \ldots, \overrightarrow{p_{i,|P_i|}}\}$. For every word in Σ^* the Parikh-vector can be encoded in binary in logarithmic space. This Parikh-vector can clearly be computed by a deterministic Turing machine that works in linear time and logarithmic space. For an $i \in I$ and a $\vec{v} \in \mathbb{N}^k$ the system $\Sigma_{j=1}^{|P_i|} \lambda_j \overrightarrow{p_{i,j}} = \vec{v} - \vec{c}_i$ of linear equations can be solved by a deterministic Turing machine that works in polynomial time and linear space in the size of $\vec{v}$ because the number of additions and multiplications of rational numbers that we have to perform can be bounded from above by a bound that only depends on k. For an $i \in I$ and a $\vec{v} \in \mathbb{N}^k$ the system $\Sigma_{j=1}^{|P_i|} \lambda_j \overrightarrow{p_{i,j}} = \vec{v} - \vec{c_i}$ has at most one solution in the rational numbers and we have $\vec{v} \in \mathsf{L}(\vec{c_i}, P_i)$ if and only if there is a solution in the natural numbers. □

We give a new simple proof of the result by Latteux that each permutation closed semirecognizable language over a binary alphabet is context free [48]:

Theorem 4.43. *Each permutation closed semilinear subset of $\{a, b\}^*$ is accepted by a counter automaton.*

Proof. Since the family of languages accepted by counter automata is closed under the operation of union and each semilinear set can be written as a finite union of linear sets with linearly independent periods by Theorem 2.22, it suffices to show that for $c, d, e, f, g, h \in \mathbb{N}$ the language

$$\psi_{\{a,b\}}^{-1}\left(\left\{ \begin{pmatrix} c \\ d \end{pmatrix} + \lambda \begin{pmatrix} e \\ f \end{pmatrix} + \mu \begin{pmatrix} g \\ h \end{pmatrix} \,\middle|\, \lambda, \mu \in \mathbb{N} \right\}\right)$$

is accepted by a counter automaton. So, let $c, d, e, f, g, h \in \mathbb{N}$ and define the maps $M, N : \{p, q, r\} \to \mathbb{N}$ through $M(p) = c$, $M(q) = e$, $M(r) = g$, $N(p) = d$, $N(q) = f$, and $N(r) = h$. Let

$$Q = \{t\} \cup \bigcup_{s \in \{p,q,r\}} \{ s_{m,n} \mid m \in \{0, 1, \ldots, M(s)\}, n \in \{0, 1, \ldots, N(s)\} \}$$

and $\Gamma = \{\bot, a, b\}$ and consider the pushdown automaton A given by

$$A = (Q, \{a, b\}, \Gamma, \delta, p_{0,0}, \bot, \{t\}).$$

The map $\delta : Q \times \{a, b, \lambda\} \times \Gamma \to 2^{Q \times \bigcup_{i=0}^{2} \Gamma^i(\Gamma^*, \Gamma, \lambda_\Gamma)}$ is given by

$$\delta = \{ ((t, y, x), \emptyset) \mid y \in \{a, b, \lambda\}, x \in \Gamma \} \cup \bigcup_{s \in \{p,q,r\}} \delta_s,$$

where for $s \in \{p, q, r\}$ the map δ_s, that maps from

$$\{ s_{m,n} \mid m \in \{0, 1, \ldots, M(s)\}, n \in \{0, 1, \ldots, N(s)\} \} \times \{a, b, \lambda\} \times \Gamma$$

to $2^{Q\times\bigcup_{i=0}^{2}\Gamma^{i}(\Gamma^*,\Gamma,\lambda_\Gamma)}$, is defined as

$$\{ ((s_{m,n}, \lambda, a), \{(s_{m+1,n}, \lambda)\})$$
$$\mid m \in \{0, 1, \ldots, M(s) - 1\}, n \in \{0, 1, \ldots, N(s)\}\}$$
$$\cup\{ ((s_{M(s),n}, \lambda, a), \emptyset) \mid n \in \{0, 1, \ldots, N(s) - 1\}\}$$
$$\cup\{ ((s_{m,n}, a, x), \{(s_{m+1,n}, x)\})$$
$$\mid x \in \Gamma, m \in \{0, 1, \ldots, M(s) - 1\}, n \in \{0, 1, \ldots, N(s)\}\}$$
$$\cup\{ ((s_{M(s),n}, a, x), \{(s_{M(s),n}, ax)\}) \mid x \in \{a, \bot\}, n \in \{0, 1, \ldots, N(s)\}\}$$
$$\cup\{ ((s_{M(s),n}, a, b), \emptyset) \mid n \in \{0, 1, \ldots, N(s)\}\}$$
$$\cup\{ ((s_{m,n}, \lambda, b), \{(s_{m,n+1}, \lambda)\})$$
$$\mid m \in \{0, 1, \ldots, M(s)\}, n \in \{0, 1, \ldots, N(s) - 1\}\}$$
$$\cup\{ ((s_{m,N(s)}, \lambda, b), \emptyset) \mid m \in \{0, 1, \ldots, M(s) - 1\}\}$$
$$\cup\{ ((s_{m,n}, b, x), \{(s_{m,n+1}, x)\})$$
$$\mid x \in \Gamma, m \in \{0, 1, \ldots, M(s)\}, n \in \{0, 1, \ldots, N(s) - 1\}\}$$
$$\cup\{ ((s_{m,N(s)}, b, x), \{(s_{m,N(s)}, bx)\}) \mid x \in \{b, \bot\}, m \in \{0, 1, \ldots, M(s)\}\}$$
$$\cup\{ ((s_{m,N(s)}, b, a), \emptyset) \mid m \in \{0, 1, \ldots, M(s)\}\}$$
$$\cup\{ ((s_{M(s),N(s)}, \lambda, x), \{(q_{0,0}, x), (r_{0,0}, x)\}) \mid x \in \{a, b\}\}$$
$$\cup\{ ((s_{M(s),N(s)}, \lambda, \bot), \{(q_{0,0}, \bot), (r_{0,0}, \bot), (t, \bot)\})\}$$
$$\cup\{ ((s_{m,n}, \lambda, \bot), \emptyset)$$
$$\mid (m, n) \in (\{0, 1, \ldots, M(s)\} \times \{0, 1, \ldots, N(s)\}) \setminus \{(M(s), N(s))\}\}.$$

For $s \in \{p, q, r\}$, $w, w' \in \{a, b\}^*$, $v, v' \in \Gamma^*$, $m, m' \in \{0, 1, \ldots, M(s)\}$, and $n, n' \in \{0, 1, \ldots, N(s)\}$ that fulfil $m' + n' \geq m + n$ and $(s_{m,n}, w, v) \vdash_A (s_{m',n'}, w', v')$ we write $(s_{m,n}, w, v) \nearrow (s_{m',n'}, w', v')$. Let $s \in \{p, q, r\}$. For $k \in \{0, 1, \ldots, M(s)\}$ it holds $(s_{0,0}, \lambda, a^k \bot) \nearrow^k (s_{k,0}, \lambda, \bot)$, while for $k \in \{0, 1, \ldots, N(s)\}$ we have $(s_{0,0}, \lambda, b^k \bot) \nearrow^k (s_{0,k}, \lambda, \bot)$. So, for $w \in$

$\{a, b\}^*$ and $k, \ell \geq 0$ we get

$$(s_{0,0}, w, a^k \bot) \nearrow^* (s_{M(s),N(s)}, \lambda, a^\ell \bot) \quad \Leftrightarrow \quad \psi_{\{a,b\}}(w) = \binom{M(s) + \ell - k}{N(s)},$$

$$(s_{0,0}, w, a^k \bot) \nearrow^* (s_{M(s),N(s)}, \lambda, b^\ell \bot) \quad \Leftrightarrow \quad \psi_{\{a,b\}}(w) = \binom{M(s) - k}{N(s) + \ell},$$

$$(s_{0,0}, w, b^k \bot) \nearrow^* (s_{M(s),N(s)}, \lambda, a^\ell \bot) \quad \Leftrightarrow \quad \psi_{\{a,b\}}(w) = \binom{M(s) + \ell}{N(s) - k},$$

$$(s_{0,0}, w, b^k \bot) \nearrow^* (s_{M(s),N(s)}, \lambda, b^\ell \bot) \quad \Leftrightarrow \quad \psi_{\{a,b\}}(w) = \binom{M(s)}{N(s) + \ell - k}.$$

For all $w \in \{a, b\}^*$, $x \in \{a, b\}$, and $k \geq 0$ with

$$\psi_{\{a,b\}}(x^k w) \in \left\{ \binom{M(s) + \ell_1}{N(s) + \ell_2} \;\middle|\; \ell_1, \ell_2 \in \mathbb{N} \right\}$$

there is a prefix v of w such that

$$\psi_{\{a,b\}}(x^k v) \in \left\{ \binom{M(s) + \ell_1}{N(s) + \ell_2} \;\middle|\; \ell_1, \ell_2 \in \mathbb{N} \wedge \ell_1 \ell_2 = 0 \right\}.$$

This gives us

$$L(A) = \psi_{\{a,b\}}^{-1} \left(\left\{ \binom{c}{d} + \lambda \binom{e}{f} + \mu \binom{g}{h} \;\middle|\; \lambda, \mu \in \mathbb{N} \right\} \right).$$

In all situations there are no a's or no b's on the stack, so A can be simulated by a counter automaton, which proves the theorem. $\qquad \square$

For $k \geq 0$ and $S \subseteq \mathbb{N}^k$ let $V_S = \mathsf{span}\left((S - S) \cap \mathbb{N}^k \right)$ and

$$X_S = \left\{ i \in \{1, 2, \ldots, k\} \;\middle|\; \exists \vec{v} \in (S - S) \cap \mathbb{N}^k : v_i > 0 \right\}.$$

Using this notation we can give an easy characterization when a strongly semirecognizable subset of $\mathbb{N}^k$ is recognizable:

Proposition 4.44. *Let $k \geq 0$ and $S \subseteq \mathbb{N}^k$ be strongly semirecognizable. Then, $S \subseteq \mathbb{N}^k$ is recognizable if and only if $\dim(V_S) = |X_S|$.*

Proof. Let $\dim(V_S) < |X_S|$. Then, there is an $i \in X_S$ with $\overrightarrow{e_{k,i}} \notin V_S$. For all $\vec{s} \in S$ and $m, n \in \mathbb{N}$ with $m \neq n$ we have $\vec{s} + m\overrightarrow{e_{k,i}} \nsim_S \vec{s} + n\overrightarrow{e_{k,i}}$. So, $S \subseteq \mathbb{N}^k$ is not recognizable. Let now $\dim(V_S) = |X_S|$ and set

$$L = \mathsf{La}\left(\vec{0}, \min\left((S - S) \cap \mathbb{N}^k \setminus \{\vec{0}\}\right)\right).$$

Then, there is an $n \in \mathbb{N}_{>0}$ such that for every $i \in X_S$ it holds $n\overrightarrow{e_{k,i}} \in L$. Thus, $L \subseteq \mathbb{N}^k$ is recognizable. By Proposition 4.28 we have $S = \min(S) + L \subseteq \mathbb{N}^k$, which is recognizable by Proposition 2.8. $\qquad\square$

Both implications of Proposition 4.44 do not hold for all semirecognizable subsets of $\mathbb{N}^k$ as the next two examples show.

Example 4.45. Let $S = \{(0,0), (1,1)\} \subset \mathbb{N}^2$, which is clearly recognizable. Then, we have $(S - S) \cap \mathbb{N}^2 = S$. This gives us $V_S = \{(a,a) \mid a \in \mathbb{R}\} \subset \mathbb{R}^2$ and $X_S = \{1, 2\}$.

Example 4.46. Let

$$S = \{(a,a) \mid a \in \mathbb{N}_{>0}\} \cup \{(1,0)\} \subset \mathbb{N}^2.$$

We have $S/\sim_S = \{[(1,1)], [(1,0)]\}$, but S is not recognizable. It holds

$$(S - S) \cap \mathbb{N}^2 = \{(a,b) \in \mathbb{N}^2 \mid b - a \in \{0,1\}\},$$

which implies $V_S = \mathbb{R}^2$ and $X_S = \{1, 2\}$.

However, a semirecognizable subset S of a monoid is recognizable if and only if all elements of $S/\sim_S$ are recognizable. Hence, we get:

Proposition 4.47. *Let $k \geq 0$ and $S \subseteq \mathbb{N}^k$ be semirecognizable. Then, the set $S \subseteq \mathbb{N}^k$ is recognizable if and only if for all $T \in S/\sim_S$ it holds $\dim(V_T) = |X_T|$.* $\qquad\square$

We investigate context-freeness of the Parikh-preimage of a semirecognizable subset of $\mathbb{N}^k$ in the following.

Proposition 4.48. *Let $k > 0$, $m \geq 0$, and $S_1, S_2, \ldots, S_m$ be strongly semirecognizable subsets of $\mathbb{N}^k$ such that for all $i, j \in \{1, 2, \ldots, m\}$ it holds $V_{S_i} = V_{S_j}$ and $\dim(V_{S_i}) = |X_{S_i}| - 1$. Then, $\psi^{-1}_{\{a_1, a_2, \ldots, a_k\}} (\bigcup_{i=1}^m S_i)$ is accepted by a λ-free deterministic counter automaton.*

Proof. Let X be the subset of $\{1, 2, \ldots, k\}$ and V be the linear subspace of $\mathbb{R}^k$ such that for all $i \in \{1, 2, \ldots, m\}$ we have $X = X_{S_i}$ and $V = V_{S_i}$. There is a $p \in X$ with $\overrightarrow{e_{k,p}} \notin V$. There exists an $n \in \mathbb{N}_{>0}$ such that for all $i \in \{1, 2, \ldots, m\}$ and $q \in X$ we have

$$n\overrightarrow{e_{k,q}} \in \mathsf{La}\left(\vec{0}, \min\left((S_i - S_i) \cap \mathbb{N}^k \setminus \left\{\vec{0}\right\}\right) \cup \{\overrightarrow{e_{k,p}}\}\right).$$

Hence,

$$\psi^{-1}_{\{a_1, a_2, \ldots, a_k\}} \left(\bigcup_{i=1}^m \left(\min(S_i) + \mathsf{La}\left(\vec{0}, \min\left((S_i - S_i) \cap \mathbb{N}^k \setminus \left\{\vec{0}\right\}\right)\right)\right)\right)$$

is accepted by a λ-free deterministic counter automaton. By Proposition 4.28 for all $i \in \{1, 2, \ldots, m\}$ it holds

$$S_i = \min(S_i) + \mathsf{La}\left(\vec{0}, \min\left((S_i - S_i) \cap \mathbb{N}^k \setminus \left\{\vec{0}\right\}\right)\right).$$

$\square$

The next lemma is needed for our study of context-freeness.

Lemma 4.49. *Let $k \geq 0$, $S \subseteq \mathbb{N}^k$, and $T_1, T_2 \in \mathbb{N}^k/{\sim_S}$ such that $(T_2 - T_1) \cap \mathbb{N}^k \neq \emptyset$. Then, it holds $(T_1 - T_1) \cap \mathbb{N}^k \subseteq T_2 - T_2$.*

Proof. Let $\vec{u}, \vec{v} \in T_1$ with $\vec{u} - \vec{v} \in \mathbb{N}^k$ and $(\vec{t_1}, \vec{t_2}) \in T_1 \times T_2$ with $\vec{t_1} \leq \vec{t_2}$. It follows $\vec{t_1} + \vec{u} - \vec{v} \in T_1$. We get $(\vec{t_2} + \vec{u} - \vec{v}) - (\vec{t_1} + \vec{u} - \vec{v}) = \vec{t_2} - \vec{t_1}$. This gives us $\vec{t_2} + \vec{u} - \vec{v} \in T_2$, which implies $\vec{u} - \vec{v} \in T_2 - T_2$. $\square$

From Proposition 4.48 and Lemma 4.49 we get:

Proposition 4.50. *Each permutation closed semirecognizable subset of $\{a, b\}^*$ is accepted by a λ-free deterministic counter automaton.* $\square$

Now we study context-freeness of arbitrary permutation closed semirecognizable languages.

Proposition 4.51. *Let $k > 0$ and $S \subseteq \mathbb{N}^k$ be semirecognizable. Then, the language $\psi^{-1}_{\{a_1,a_2,\ldots,a_k\}}(S)$ is context free if and only if for all $T \in S/\sim_S$ it holds $\dim(V_T) \geq |X_T| - 1$. In this case, $\psi^{-1}_{\{a_1,a_2,\ldots,a_k\}}(S)$ is accepted by a counter automaton.*

Proof. If for all $T \in S/\sim_S$ it holds $\dim(V_T) \geq |X_T| - 1$, $\psi^{-1}_{\{a_1,a_2,\ldots,a_k\}}(S)$ is accepted by a counter automaton by Proposition 4.48. Assume now that there is a $T \in S/\sim_S$ with $\dim(V_T) \leq |X_T| - 2$. There are $i, j \in X_T$ with $i \neq j$ such that $\mathsf{span}\left(\{\overrightarrow{e_{k,i}}, \overrightarrow{e_{k,j}}\}\right) \cap V_T = \{\vec{0}\}$. Let $\vec{t} \in T$ and n' be the maximum of

$$\left\{ |\lambda|, |\mu| \;\middle|\; \lambda, \mu \in \mathbb{Z} \wedge \exists U \in S/\sim_S : \vec{t} + \lambda\overrightarrow{e_{k,i}} + \mu\overrightarrow{e_{k,j}} \in \min(U) + V_T \right\}.$$

Set $n = n' + 1$. There is an $\vec{x} \in T - T$ with $n\overrightarrow{e_{k,i}} + n\overrightarrow{e_{k,j}} \leq \vec{x}$. Let $\vec{s} \in S$ with $\vec{s} - \vec{t} \in \mathsf{La}(\vec{0}, \{\vec{x}, \overrightarrow{e_{k,i}}, \overrightarrow{e_{k,j}}\})$. Furthermore, let $U \in S/\sim_S$ with $\vec{s} \in U$. By Lemma 4.49 we have $(U - U) \cap \mathbb{N}^k \subseteq T - T$. Let $\vec{u} \in \min(U)$ with $\vec{u} \leq \vec{s}$. Then, $\vec{s} - \vec{t} = (\vec{u} - \vec{t}) + (\vec{s} - \vec{u})$ is in

$$V_T + \left\{ \lambda\overrightarrow{e_{k,i}} + \mu\overrightarrow{e_{k,j}} \mid \lambda, \mu \in \{-n+1, -n+2, \ldots, n-1\} \right\}.$$

Hence, for any monoid homomorphism $h : \{a, b, c\}^* \to \{a_1, a_2, \ldots, a_k\}^*$ with $h(a) \in \psi^{-1}_{\{a_1,a_2,\ldots,a_k\}}(\{n\overrightarrow{e_{k,i}}\})$, $h(b) \in \psi^{-1}_{\{a_1,a_2,\ldots,a_k\}}(\{n\overrightarrow{e_{k,j}}\})$, and

$$h(c) \in \psi^{-1}_{\{a_1,a_2,\ldots,a_k\}}(\{\vec{x} - n\overrightarrow{e_{k,i}} - n\overrightarrow{e_{k,j}}\}),$$

we get

$$h^{-1}\left(\psi^{-1}_{\{a_1,a_2,\ldots,a_k\}}\left(\left\{\vec{s} - \vec{t} \;\middle|\; \vec{s} \in S \wedge \vec{t} \leq \vec{s}\right\}\right)\right)$$
$$= \left\{ w \in \{a, b, c\}^* \mid |w|_a = |w|_b = |w|_c \right\}.$$

Since the family **CF** is closed under the operation of inverse homomorphism, the language $\psi^{-1}_{\{a_1,a_2,\ldots,a_k\}}\left(\left\{\vec{s} - \vec{t} \mid \vec{s} \in S \wedge \vec{t} \leq \vec{s}\right\}\right)$ is not context free. This implies that S is not context free, as well. $\square$

Example 4.52. Let $n > 0$ and L_n be the permutation closed language

$$\{\, w \in \{a,b,c\}^* \mid |w|_a = |w|_b \leq |w|_c < |w|_a + n \,\}.$$

For $v, w \in L_n$ we have $v \sim_{L_n} w$ if and only if $|v|_c - |v|_a = |w|_c - |w|_a$. Thus, it holds $|L_n/{\sim_{L_n}}| = n$, which tells us that L_n is n-semirecognizable, but not $(n-1)$-semirecognizable. Let $T = \{\, (a,a,a) \mid a \in \mathbb{N} \,\} \subset \mathbb{N}^3$. We have $T \in \psi(L_n)/{\sim_{\psi(L_n)}}$. Furthermore, it holds $(T - T) \cap \mathbb{N}^3 = T$. This gives us $V_T = \{\, (a,a,a) \mid a \in \mathbb{R} \,\} \subset \mathbb{R}^3$ and $X_T = \{1,2,3\}$. So, L_n is not context free.

Propositions 4.48 and 4.51 give us:

Proposition 4.53. *Every permutation closed, strongly semirecognizable, context-free language is accepted by a λ-free deterministic counter automaton.* $\qquad\square$

Propositions 4.48 and 4.51 and Lemma 4.49 imply:

Proposition 4.54. *Each permutation closed, semirecognizable, context-free sublanguage of $\{a,b,c\}^*$ is accepted by a λ-free deterministic counter automaton.* $\qquad\square$

To study deterministic context-freeness of the Parikh-preimage of a semirecognizable subset of $\mathbb{N}^k$, we need the following notation. Let $k \geq 0$, $S \subseteq \mathbb{N}^k$ with $\dim(V_S) = |X_S| - 1$, and $\vec{w} \in \mathbb{N}^k \setminus V_S$ such that for all $i \in \{1, 2, \ldots, k\} \setminus X_S$ it holds $w_i = 0$. Then, for every $i \in X_S$ there are unique $\overrightarrow{x_{S,\vec{w},i}} \in V_S$ and $\lambda_{S,\vec{w},i} \in \mathbb{Q}$ with $\overrightarrow{e_{k,i}} = \overrightarrow{x_{S,\vec{w},i}} + \lambda_{S,\vec{w},i}\vec{w}$. Set

$$\mu_{S,\vec{w}} = \min\left(\{\, \nu \in \mathbb{Q}_{>0} \mid \forall i \in X_S : \nu\lambda_{S,\vec{w},i} \in \mathbb{Z} \,\}\right)$$

and let $\overrightarrow{v_{S,\vec{w}}} \in \mathbb{Z}^k$ be the vector that obeys $\pi_{k,\{1,2,\ldots,k\}\setminus X_S}(\overrightarrow{v_{S,\vec{w}}}) = \vec{0}$ and for every $i \in X_S$ fulfils $\pi_{k,\{i\}}(\overrightarrow{v_{S,\vec{w}}}) = \mu_{S,\vec{w}}\lambda_{S,\vec{w},i}$. Let $\vec{y} \in \mathbb{N}^k \setminus V_S$ such that for all $i \in \{1, 2, \ldots, k\} \setminus X_S$ it holds $y_i = 0$. Then, it is easy to see that $\overrightarrow{v_{S,\vec{y}}} \in \{\overrightarrow{v_{S,\vec{w}}}, -\overrightarrow{v_{S,\vec{w}}}\}$. We set $i_S = \min(\{\, i \in X_S \mid \overrightarrow{e_{k,i}} \notin V_S \,\})$ and $\vec{v_S} = \overrightarrow{v_{S,\overrightarrow{e_{k,i_S}}}}$. It holds $\pi_{k,\{i_S\}}(\vec{v_S}) > 0$. There is a $\vec{z} \in (S - S) \cap \mathbb{N}^k$ with $z_{i_S} > 0$. This gives us $\vec{z} - \overrightarrow{e_{k,i_S}} \in \mathbb{N}^k$. So, there is an $i \in X_S$ with $\pi_{k,\{i\}}(\vec{v_S}) < 0$. This shows that $\vec{v_S}$ has at least one positive and at least one negative component.

Proposition 4.55. *Let $k > 0$ and $S \subseteq \mathbb{N}^k$ be semirecognizable such that for all $T \in S/\sim_S$ it holds $\dim(V_T) \geq |X_T| - 1$. Then, $\psi_{\{a_1, a_2, \ldots, a_k\}}^{-1}(S)$ is deterministic context free if and only if for all sets $T, U \in S/\sim_S$ with $\dim(V_T) = |X_T| - 1$ and $\dim(V_U) = |X_U| - 1$ at least one of the following two conditions is true.*

- *The set $\pi_{k, X_T \cap X_U}(\{\vec{v_T}, \vec{v_U}\})$ is collinear.*

- *For all $i, j \in X_T \cap X_U$ it holds $\pi_{k,i}(\vec{v_T}) \cdot \pi_{k,j}(\vec{v_T}) \geq 0$ and $\pi_{k,i}(\vec{v_U}) \cdot \pi_{k,j}(\vec{v_U}) \geq 0$.*

Proof. Set $\Sigma = \{a_1, a_2, \ldots, a_k\}$. Given the proof of Proposition 4.51, it is not hard to see that if for all $T, U \in S/\sim_S$ with $\dim(V_T) = |X_T| - 1$ and $\dim(V_U) = |X_U| - 1$ at least one of the two conditions from the lemma is true, then $\psi_\Sigma^{-1}(S)$ is accepted by a DPDA. Assume that $\psi_\Sigma^{-1}(S)$ is deterministic context free and that there are $T, U \in S/\sim_S$ with $\dim(V_T) = |X_T| - 1$ and $\dim(V_U) = |X_U| - 1$ such that $\pi_{k, X_T \cap X_U}(\{\vec{v_T}, \vec{v_U}\})$ is linearly independent and there are $i, j \in X_T \cap X_U$ with $\pi_{k,i}(\vec{v_T}) \cdot \pi_{k,j}(\vec{v_T}) < 0$. If the set $\pi_{k, \{i,j\}}(\{\vec{v_T}, \vec{v_U}\})$ is linearly independent, set $\ell = i$. Otherwise, let $\ell \in X_T \cap X_U$ such that the set $\pi_{k, \{\ell, j\}}(\{\vec{v_T}, \vec{v_U}\})$ is linearly independent. There is a $p \in X_U$ and a $q \in \{\ell, j\}$ with $\pi_{k,p}(\vec{v_U}) \cdot \pi_{k,q}(\vec{v_U}) < 0$. By Lemma 4.49 there is an $M \in \mathbb{N}_{>0}$ such that for all $W \in \{T, U\}$ we have

$$(V_W + \{\, m\overrightarrow{e_{k,i_W}} \mid m \in \mathbb{R} \wedge |m| \geq M \,\}) \cap (S - W) \cap \mathbb{N}^k = \emptyset.$$

Let $w_T \in \psi_\Sigma^{-1}(T) \setminus \{\lambda\}$ and $w_U \in \psi_\Sigma^{-1}(U) \setminus \{\lambda\}$ such that the last symbol of w_T and the last symbol of w_U are different. Set

$$L = \{\, wa_i \mid w \in \Sigma^* \wedge wa_i w_T \in \psi_\Sigma^{-1}(S) \,\}$$
$$\cup \{\, wa_j \mid w \in \Sigma^* \wedge wa_j w_U \in \psi_\Sigma^{-1}(S) \,\}.$$

The language L is deterministic context free because **DCF** is closed under intersection with a regular language and under right quotient with a regular language, see the book *Introduction to Automata Theory, Languages, and Computation* by Hopcroft and Ullman [39]. Let n be the constant that Lemma 2.15 gives us for L. There are numbers $r \in \{p, q\}$ and $n_1, n_2, n_3, n_4, n_5 \in \mathbb{N}$ with $\min(n_2, n_3) > n$ and $n_5 > 0$ such that

$$a_j^{n_1} a_\ell^{n_2} a_i^{n_3} w_T \in \psi_\Sigma^{-1}(T) \quad \text{and} \quad a_j^{n_1} a_\ell^{n_2} a_i^{n_3} a_r^{n_4} a_j^{n_5} w_U \in \psi_\Sigma^{-1}(U).$$

Set $\sigma = a_i$, $x = a_j^{n_1} a_\ell^{n_2} a_i^{n_3-1}$, $y = \lambda$, and $z = a_r^{n_4} a_j^{n_5}$. Then, Lemma 2.15 gives a contradiction. $\qquad\square$

Example 4.56. Let $n \geq 2$ and L_n be the permutation closed language

$$\{w \in \{a,b,c,d\}^* \mid (|w|_a \leq |w|_b < |w|_a + n - 1 \wedge |w|_c = 0 < |w|_d)$$
$$\vee (|w|_a = 2 \cdot |w|_b \wedge |w|_d = 0 < |w|_c)\}.$$

For $v, w \in L_n$ we have $v \sim_{L_n} w$ if and only if

$$(|v|_c = |w|_c = 0 \wedge |v|_b - |v|_a = |w|_b - |w|_a) \vee |v|_d = |w|_d = 0.$$

Hence, we get $|L_n/\sim_{L_n}| = n$. There are $n - 1$ elements $T \in \psi(L_n)/\sim_{\psi(L_n)}$ with

$$(T - T) \cap \mathbb{N}^4 = \{(a,a,0,b) \mid a,b \in \mathbb{N}\}$$

and one $U \in \psi(L_n)/\sim_{\psi(L_n)}$ with

$$(U - U) \cap \mathbb{N}^4 = \{(2a,a,b,0) \mid a,b \in \mathbb{N}\}.$$

Let $T, U \in \psi(L_n)/\sim_{\psi(L_n)}$ with the properties from the last sentence. It holds $\dim(V_T) = \dim(V_U) = 2$ and $|X_T| = |X_U| = 3$. By Proposition 4.51 the language L_n is accepted by a counter automaton. We have $i_T = 1$ and $(0,1,0,0) = (1,1,0,0) - (1,0,0,0)$. This gives us $v_T = (1,-1,0,0)$. Furthermore, it holds $i_U = 1$ and $(0,1,0,0) = (2,1,0,0) - (2,0,0,0)$. That implies $v_U = (1,-2,0,0)$. By Proposition 4.55 the language L_n is not deterministic context free.

We investigate acceptance by a λ-free deterministic pushdown automaton:

Proposition 4.57. *Let $k > 0$ and $S \subseteq \mathbb{N}^k$ be semirecognizable such that for all $T \in S/\sim_S$ it holds $\dim(V_T) \geq |X_T| - 1$. Then, $\psi_{\{a_1,a_2,\ldots,a_k\}}^{-1}(S)$ is accepted by a λ-free deterministic pushdown automaton if and only if for all $T, U \in S/\sim_S$ with $\dim(V_T) = |X_T| - 1$ and $\dim(V_U) = |X_U| - 1$ at least one of the following two conditions is true.*

- *The set $\pi_{k,X_T \cap X_U}(\{\vec{v_T}, \vec{v_U}\})$ is collinear.*

- *The following two conditions are both true.*

- For all $i, j \in X_T \cap X_U$ it holds $\pi_{k,i}(\vec{v_T}) \cdot \pi_{k,j}(\vec{v_T}) \geq 0$ and also $\pi_{k,i}(\vec{v_U}) \cdot \pi_{k,j}(\vec{v_U}) \geq 0$.

- For all $i \in X_T \cap X_U$ we have $\pi_{k,i}(\vec{v_T}) = 0$ if and only if $\pi_{k,i}(\vec{v_U}) = 0$.

Proof. Set $\Sigma = \{a_1, a_2, \ldots, a_k\}$. It is not hard to see that if for all $T, U \in S/\sim_S$ with $\dim(V_T) = |X_T| - 1$ and $\dim(V_U) = |X_U| - 1$ at least one of the two conditions from the lemma is true, then $\psi_\Sigma^{-1}(S)$ is accepted by a λ-free deterministic pushdown automaton. Assume that there are $T, U \in S/\sim_S$ with $\dim(V_T) = |X_T| - 1$ and $\dim(V_U) = |X_U| - 1$ such that the set $\pi_{k,X_T \cap X_U}(\{\vec{v_T}, \vec{v_U}\})$ is linearly independent, for all $i, j \in X_T \cap X_U$ it holds $\pi_{k,i}(\vec{v_T}) \cdot \pi_{k,j}(\vec{v_T}) \geq 0$ and $\pi_{k,i}(\vec{v_U}) \cdot \pi_{k,j}(\vec{v_U}) \geq 0$, and there is an $\ell \in X_T \cap X_U$ with $\pi_{k,\ell}(\vec{v_T}) = 0$ and $\pi_{k,\ell}(\vec{v_U}) \neq 0$. Let $p \in X_T \cap X_U$ such that $\pi_{k,\{p,\ell\}}(\{\vec{v_T}, \vec{v_U}\})$ is linearly independent. Furthermore, let $q \in X_T$ and $r \in X_U$ with $\pi_{k,q}(\vec{v_T}) \cdot \pi_{k,p}(\vec{v_T}) < 0$ and $\pi_{k,r}(\vec{v_U}) \cdot \pi_{k,\ell}(\vec{v_U}) < 0$. By Lemma 4.49 there is an $M \in \mathbb{N}_{>0}$ such that

$$\emptyset = (V_T + \{ m\overrightarrow{e_{k,p}} \mid m \in \mathbb{R} \wedge |m| \geq M \}) \cap (S - T) \cap \mathbb{N}^k$$
$$= (V_U + \{ m\overrightarrow{e_{k,\ell}} \mid m \in \mathbb{R} \wedge |m| \geq M \}) \cap (S - U) \cap \mathbb{N}^k.$$

Let $N \in \mathbb{N}_{>0}$ such that for all $w_T \in \psi_\Sigma^{-1}(T)$, $w_U \in \psi_\Sigma^{-1}(U)$, and numbers $n_1, n_2, n_3, n_4 \in \mathbb{N}$ it holds

$$a_p^{|\pi_{k,q}(\vec{v_T})| \cdot n_1 N} a_\ell^{n_2 N} a_q^{|\pi_{k,p}(\vec{v_T})| \cdot n_1 N} w_T \in \psi_\Sigma^{-1}(T)$$

and

$$a_p^{|\pi_{k,r}(\vec{v_U})| \cdot n_3 N} a_\ell^{|\pi_{k,r}(\vec{v_U})| \cdot n_4 N} a_r^{(|\pi_{k,p}(\vec{v_U})| \cdot n_3 + |\pi_{k,\ell}(\vec{v_U})| \cdot n_4)N} w_U \in \psi_\Sigma^{-1}(U).$$

Let $w_T \in \psi_\Sigma^{-1}(T)$ and define $f : \mathbb{N} \to \mathbb{N}$ through

$$f(n) = |\pi_{k,p}(\vec{v_T})| \cdot |\pi_{k,r}(\vec{v_U})| \cdot nMN + |w_T|.$$

For all $i, j \in \mathbb{N}_{>0}$ let

$$x_i = a_p^{|\pi_{k,q}(\vec{v_T})| \cdot |\pi_{k,r}(\vec{v_U})| \cdot iMN},$$
$$y_{i,j} = a_\ell^{|\pi_{k,r}(\vec{v_U})| \cdot jMN},$$
$$w_{i,j} = a_q^{|\pi_{k,p}(\vec{v_T})| \cdot |\pi_{k,r}(\vec{v_U})| \cdot iMN} w_T.$$

By Lemma 2.17 the language $\psi^{-1}_{\{a_1,a_2,\ldots,a_k\}}(S)$ is not accepted by a λ-free deterministic pushdown automaton. $\qquad\qquad\square$

Example 4.58. Let $n \geq 2$ and L_n be the permutation closed language

$$\{w \in \{a,b,c,d\}^*$$
$$\mid |w|_d = 0 < |w|_a \leq |w|_c < |w|_a + n - 1 \vee |w|_c = 0 < |w|_b = |w|_d\}.$$

For $v, w \in L_n$ we have $v \sim_{L_n} w$ if and only if

$$(|v|_d = |w|_d = 0 \wedge |v|_c - |v|_a = |w|_c - |w|_a) \vee |v|_c = |w|_c = 0.$$

Hence, we get $|L_n/\!\!\sim_{L_n}| = n$. There are $n - 1$ elements $T \in \psi(L_n)/\!\!\sim_{\psi(L_n)}$ with

$$(T - T) \cap \mathbb{N}^4 = \{(a,b,a,0) \mid a, b \in \mathbb{N}\}$$

and one $U \in \psi(L_n)/\!\!\sim_{\psi(L_n)}$ with

$$(U - U) \cap \mathbb{N}^4 = \{(a,b,0,b) \mid a, b \in \mathbb{N}\}.$$

Let $T, U \in \psi(L_n)/\!\!\sim_{\psi(L_n)}$ with the properties from the last sentence. It holds $\dim(V_T) = \dim(V_U) = 2$, $X_T = \{1,2,3\}$, and $X_U = \{1,2,4\}$. We have $i_T = 1$ and $(0,0,1,0) = (1,0,1,0) - (1,0,0,0)$. This gives us $v_T = (1,0,-1,0)$. It holds $i_U = 2$ and $(0,0,0,1) = (0,1,0,1) - (0,1,0,0)$. That implies $v_U = (0,1,0,-1)$. By Proposition 4.55 the language L_n is deterministic context free. On the other hand, L_n is not accepted by a λ-free deterministic pushdown automaton by Proposition 4.57.

We turn to acceptance by a deterministic counter automaton:

Proposition 4.59. *Let $k > 0$ and $S \subseteq \mathbb{N}^k$ be semirecognizable such that for all $T \in S/\!\!\sim_S$ it holds $\dim(V_T) \geq |X_T| - 1$. Then, $\psi^{-1}_{\{a_1,a_2,\ldots,a_k\}}(S)$ is accepted by a deterministic counter automaton if and only if for all $T, U \in S/\!\!\sim_S$ with $\dim(V_T) = |X_T| - 1$ and $\dim(V_U) = |X_U| - 1$ the set $\pi_{k,X_T \cap X_U}(\{\vec{v_T}, \vec{v_U}\})$ is collinear.*

Proof. Set $\Sigma = \{a_1, a_2, \ldots, a_k\}$. It is not hard to see that if for all $T, U \in S/\!\!\sim_S$ that fulfil $\dim(V_T) = |X_T| - 1$ and $\dim(V_U) = |X_U| - 1$ the set $\pi_{k,X_T \cap X_U}(\{\vec{v_T}, \vec{v_U}\})$ is collinear, then $\psi_\Sigma^{-1}(S)$ is accepted by a deterministic counter automaton. Assume that there are $T, U \in S/\!\!\sim_S$ with $\dim(V_T) = |X_T| - 1$ and $\dim(V_U) = |X_U| - 1$ such that $\pi_{k,X_T \cap X_U}(\{\vec{v_T}, \vec{v_U}\})$ is linearly independent and for all $i, j \in X_T \cap X_U$ it holds $\pi_{k,i}(\vec{v_T}) \cdot \pi_{k,j}(\vec{v_T}) \geq 0$ and $\pi_{k,i}(\vec{v_U}) \cdot \pi_{k,j}(\vec{v_U}) \geq 0$. Let $i, j \in X_T \cap X_U$ such that $\pi_{k,\{i,j\}}(\{\vec{v_T}, \vec{v_U}\})$ is linearly independent and it holds $\pi_{k,i}(\vec{v_T}) \neq 0 \neq \pi_{k,j}(\vec{v_U})$. Furthermore, let $p \in X_T$ and $q \in X_U$ such that $\pi_{k,p}(\vec{v_T}) \cdot \pi_{k,i}(\vec{v_T}) < 0$ and also $\pi_{k,q}(\vec{v_U}) \cdot \pi_{k,j}(\vec{v_U}) < 0$. By Lemma 4.49 there is an $M \in \mathbb{N}_{>0}$ such that

$$\emptyset = (V_T + \{ m\overrightarrow{e_{k,p}} \mid m \in \mathbb{R} \wedge |m| \geq M \}) \cap (S - T) \cap \mathbb{N}^k$$
$$= (V_U + \{ m\overrightarrow{e_{k,q}} \mid m \in \mathbb{R} \wedge |m| \geq M \}) \cap (S - U) \cap \mathbb{N}^k.$$

Let $N \in \mathbb{N}_{>0}$ such that for all $w_T \in \psi_\Sigma^{-1}(T)$, $w_U \in \psi_\Sigma^{-1}(U)$, and numbers $n_1, n_2, n_3, n_4 \in \mathbb{N}$ it holds

$$a_i^{|\pi_{k,p}(\vec{v_T})| \cdot n_1 N} a_j^{|\pi_{k,p}(\vec{v_T})| \cdot n_2 N} a_p^{(|\pi_{k,i}(\vec{v_T})| \cdot n_1 + |\pi_{k,j}(\vec{v_T})| \cdot n_2)N} w_T \in \psi_\Sigma^{-1}(T)$$

and

$$a_i^{|\pi_{k,q}(\vec{v_U})| \cdot n_3 N} a_j^{|\pi_{k,q}(\vec{v_U})| \cdot n_4 N} a_q^{(|\pi_{k,i}(\vec{v_U})| \cdot n_3 + |\pi_{k,j}(\vec{v_U})| \cdot n_4)N} w_U \in \psi_\Sigma^{-1}(U).$$

Set $x = |\pi_{k,p}(\vec{v_T})| \cdot |\pi_{k,q}(\vec{v_U})| \cdot MN$. Assume that there is a deterministic counter automaton

$$C = (Q, \Sigma, \Gamma = \{\bot, A\}, \delta, s_0, \bot, F)$$

with $L(C) = \psi_\Sigma^{-1}(S)$. Define $f : \mathbb{N}^2 \to Q \times \mathbb{N}$ in the following way. For all $(n, m) \in \mathbb{N}^2$ there is a unique pair $f(n, m) = (r, \ell) \in Q \times \mathbb{N}$ such that $\left(s_0, a_i^{nx} a_j^{mx}, \bot\right) \vdash^* (r, \lambda, A^\ell \bot)$ and there is no configuration c of C with $(r, \lambda, A^\ell \bot) \vdash c$. It follows from what is written above that f is injective. Let $f_1 : \mathbb{N}^2 \to Q$ and $f_2 : \mathbb{N}^2 \to \mathbb{N}$ be the functions that fulfil $f = (f_1, f_2)$. Set

$$y = \max\left(\{ n \in \mathbb{N} \mid \exists r \in Q : (s_0, a_i^n, \bot) \vdash^* (r, \lambda, \bot) \}\right).$$

There are $z_1, z_2 \in \mathbb{N}$ with $y < z_1 < z_2$ such that $f_1(z_1, 0) = f_1(z_2, 0)$. It follows $f_2(z_1, 0) < f_2(z_2, 0)$. Moreover, for all $n \in \mathbb{N}$ we have

$$f(z_1 + n(z_2 - z_1), 0) = (f_1(z_1, 0), f_2(z_1, 0) + n(f_2(z_2, 0) - f_2(z_1, 0))).$$

113

Let u be the maximum of

$$\left\{ n \in \mathbb{N} \;\middle|\; \exists r \in Q, m \in \mathbb{N} : \left(s_0, a_i^{(z_1 + n(z_2 - z_1))x} a_j^m, \bot\right) \vdash^* (r, \lambda, \bot) \right\}.$$

There are $t_1, t_2 \in \mathbb{N}$ with $t_1 < t_2$ such that

$$f_1(z_1 + (u+1)(z_2 - z_1), t_1) = f_1(z_1 + (u+1)(z_2 - z_1), t_2).$$

This implies

$$f_2(z_1 + (u+1)(z_2 - z_1), t_1) < f_2(z_1 + (u+1)(z_2 - z_1), t_2).$$

For all $n, m \in \mathbb{N}$ we get

$$f_1(z_1 + (u+1+m)(z_2 - z_1), t_1 + n(t_2 - t_1)) = f_1(z_1 + (u+1)(z_2 - z_1), t_1)$$

and

$$\begin{aligned}
f_2(z_1 &+ (u+1+m)(z_2 - z_1), t_1 + n(t_2 - t_1)) \\
&= f_2(z_1 + (u+1+m)(z_2 - z_1), t_1) \\
&\quad + n(f_2(z_1 + (u+1)(z_2 - z_1), t_2) - f_2(z_1 + (u+1)(z_2 - z_1), t_1)),
\end{aligned}$$

where

$$\begin{aligned}
f_2(z_1 &+ (u+1+m)(z_2 - z_1), t_1) \\
&= f_2(z_1 + (u+1+m)(z_2 - z_1), 0) + f_2(z_1 + (u+1)(z_2 - z_1), t_1) \\
&\quad - f_2(z_1 + (u+1)(z_2 - z_1), 0) \\
&= f_2(z_1 + (u+1)(z_2 - z_1), t_1) + m(f_2(z_2, 0) - f_2(z_1, 0)).
\end{aligned}$$

Hence, we have

$$\begin{aligned}
f(z_1 &+ (u+1+ f_2(z_1 + (u+1)(z_2 - z_1), t_2) \\
&- f_2(z_1 + (u+1)(z_2 - z_1), t_1))(z_2 - z_1), t_1) \\
&= f(z_1 + (u+1)(z_2 - z_1), t_1 + (f_2(z_2, 0) - f_2(z_1, 0))(t_2 - t_1)),
\end{aligned}$$

a contradiction. $\qquad\square$

Propositions 4.57 and 4.59 give us:

Proposition 4.60. *Every permutation closed semirecognizable language that is accepted by a deterministic counter automaton is also accepted by a λ-free deterministic pushdown automaton.* $\qquad\square$

Example 4.61. Let $n \geq 2$ and L_n be the permutation closed language

$$\{w \in \{a,b,c,d\}^* \mid |w|_d = 0 < |w|_a + |w|_b \leq |w|_c < |w|_a + |w|_b + n - 1$$
$$\vee |w|_c = 0 < |w|_a + 2 \cdot |w|_b = |w|_d\}.$$

For $v, w \in L_n$ we have $v \sim_{L_n} w$ if and only if

$$(|v|_d = |w|_d = 0 \wedge |v|_c - |v|_a - |v|_b = |w|_c - |w|_a - |w|_b) \vee |v|_c = |w|_c = 0.$$

Hence, we get $|L_n/\sim_{L_n}| = n$. There are $n - 1$ elements $T \in \psi(L_n)/\sim_{\psi(L_n)}$ with

$$(T - T) \cap \mathbb{N}^4 = \{(a, b, a + b, 0) \mid a, b \in \mathbb{N}\}$$

and one $U \in \psi(L_n)/\sim_{\psi(L_n)}$ with

$$(U - U) \cap \mathbb{N}^4 = \{(a, b, 0, a + 2b) \mid a, b \in \mathbb{N}\}.$$

Let $T, U \in \psi(L_n)/\sim_{\psi(L_n)}$ with the properties from the last sentence. It holds $\dim(V_T) = \dim(V_U) = 2$, $X_T = \{1, 2, 3\}$, and $X_U = \{1, 2, 4\}$. We have $i_T = 1$, $(0, 1, 0, 0) = (-1, 1, 0, 0) + (1, 0, 0, 0)$, and $(0, 0, 1, 0) = (1, 0, 1, 0) - (1, 0, 0, 0)$. This gives us $v_T = (1, 1, -1, 0)$. It holds $i_U = 1$, $(0, 1, 0, 0) = (-2, 1, 0, 0) + (2, 0, 0, 0)$, and also $(0, 0, 0, 1) = (1, 0, 0, 1) - (1, 0, 0, 0)$. That implies $v_U = (1, 2, 0, -1)$. By Proposition 4.57 the language L_n is accepted by a λ-free deterministic pushdown automaton. On the other hand, L_n is not accepted by a deterministic counter automaton by Proposition 4.59.

Concerning acceptance by a λ-free deterministic counter automaton, we get:

Proposition 4.62. *Let $k > 0$ and $S \subseteq \mathbb{N}^k$ be semirecognizable such that for all $T \in S/\sim_S$ it holds $\dim(V_T) \geq |X_T| - 1$. Then, $\psi_{\{a_1,a_2,\ldots,a_k\}}^{-1}(S)$ is accepted by a λ-free deterministic counter automaton if and only if for all $T, U \in S/\sim_S$ with $\dim(V_T) = |X_T| - 1$ and $\dim(V_U) = |X_U| - 1$ we have $\mathsf{span}\left(\pi_{k, X_T \cap X_U}(\vec{v_T})\right) = \mathsf{span}\left(\pi_{k, X_T \cap X_U}(\vec{v_U})\right)$.*

Proof. Set $\Sigma = \{a_1, a_2, \ldots, a_k\}$. It is not hard to see that if for all $T, U \in S/\sim_S$ that fulfil $\dim(V_T) = |X_T| - 1$ and $\dim(V_U) = |X_U| - 1$ it holds that $\mathsf{span}\,(\pi_{k, X_T \cap X_U}(\vec{v_T})) = \mathsf{span}\,(\pi_{k, X_T \cap X_U}(\vec{v_U}))$, then $\psi_\Sigma^{-1}(S)$ is accepted by a λ-free deterministic counter automaton. Assume that there are $T, U \in S/\sim_S$ fulfilling $\dim(V_T) = |X_T| - 1$, $\dim(V_U) = |X_U| - 1$, and

$$\pi_{k, X_T \cap X_U}(\vec{v_T}) \neq \vec{0} = \pi_{k, X_T \cap X_U}(\vec{v_U}).$$

Let $i \in X_T \cap X_U$ with $\pi_{k,i}(\vec{v_T}) \neq 0$. Moreover, let $j \in X_T$ and $p, q \in X_U$ such that we have $\pi_{k,i}(\vec{v_T}) \cdot \pi_{k,j}(\vec{v_T}) < 0$ and also $\pi_{k,p}(\vec{v_U}) \cdot \pi_{k,q}(\vec{v_U}) < 0$. By Lemma 4.49 there is an $M \in \mathbb{N}_{>0}$ such that

$$\emptyset = (V_T + \{\, m\vec{e_{k,i}} \mid m \in \mathbb{R} \wedge |m| \geq M \,\}) \cap (S - T) \cap \mathbb{N}^k$$
$$= (V_U + \{\, m\vec{e_{k,p}} \mid m \in \mathbb{R} \wedge |m| \geq M \,\}) \cap (S - U) \cap \mathbb{N}^k.$$

Let $N \in \mathbb{N}_{>0}$ such that for all $w_T \in \psi_\Sigma^{-1}(T)$, $w_U \in \psi_\Sigma^{-1}(U)$, and numbers $n_1, n_2, n_3 \in \mathbb{N}$ it holds

$$a_i^{|\pi_{k,j}(\vec{v_T})| \cdot n_1 N} a_j^{|\pi_{k,i}(\vec{v_T})| \cdot n_1 N} w_T \in \psi_\Sigma^{-1}(T)$$

and

$$a_i^{n_2 N} a_p^{|\pi_{k,q}(\vec{v_U})| \cdot n_3 N} a_q^{|\pi_{k,p}(\vec{v_U})| \cdot n_3 N} w_U \in \psi_\Sigma^{-1}(U).$$

Set $x = |\pi_{k,j}(\vec{v_T})| \cdot |\pi_{k,q}(\vec{v_U})| \cdot MN$. Assume that there is a λ-free deterministic counter automaton

$$C = (Q, \Sigma, \Gamma = \{\bot, A\}, \delta, s_0, \bot, F)$$

with $L(C) = \psi_\Sigma^{-1}(S)$. Define $f : \mathbb{N}^2 \to Q \times \mathbb{N}$ in the following way. For all $(n, m) \in \mathbb{N}^2$ there is a unique pair $f(n, m) = (r, \ell) \in Q \times \mathbb{N}$ such that $\left(s_0, a_i^{nx} a_p^{mx}, \bot\right) \vdash^* (r, \lambda, A^\ell \bot)$. It follows from what is written above that for all $n_1, n_2, n_3, n_4 \in \mathbb{N}$ with $n_1 < n_2$ we have $f(n_1, 0) \neq f(n_2, 0)$ and $f(n_3, n_1) \neq f(n_4, n_2)$. Let $f_1 : \mathbb{N}^2 \to Q$ and $f_2 : \mathbb{N}^2 \to \mathbb{N}$ be the functions that fulfil $f = (f_1, f_2)$. Set

$$y = \max\left(\{\, n \in \mathbb{N} \mid \exists r \in Q : (s_0, a_i^n, \bot) \vdash^* (r, \lambda, \bot) \,\}\right).$$

There are $z_1, z_2 \in \mathbb{N}$ with $y < z_1 < z_2$ such that $f_1(z_1, 0) = f_1(z_2, 0)$. It follows $f_2(z_1, 0) < f_2(z_2, 0)$. Moreover, for all $n \in \mathbb{N}$ we have

$$f(z_1 + n(z_2 - z_1), 0) = (f_1(z_1, 0), f_2(z_1, 0) + n(f_2(z_2, 0) - f_2(z_1, 0))).$$

There are $t_1, t_2 \in \mathbb{N}$ with $t_1 < t_2 \leq |Q|$ such that

$$f_1(z_1 + |Q| \cdot x(z_2 - z_1), t_1) = f_1(z_1 + |Q| \cdot x(z_2 - z_1), t_2).$$

This implies

$$f_2(z_1 + |Q| \cdot x(z_2 - z_1), t_2) \neq f_2(z_1 + |Q| \cdot x(z_2 - z_1), t_1).$$

For all $n, m \in \mathbb{N}$ such that

$$f_2(z_1 + |Q| \cdot x(z_2 - z_1), t_2) > f_2(z_1 + |Q| \cdot x(z_2 - z_1), t_1)$$

or

$$n(f_2(z_1 + |Q| \cdot x(z_2 - z_1), t_1) - f_2(z_1 + |Q| \cdot x(z_2 - z_1), t_2))$$
$$\leq m(f_2(z_2, 0) - f_2(z_1, 0))$$

we get

$$f_1(z_1 + (|Q| \cdot x + m)(z_2 - z_1), t_1 + n(t_2 - t_1)) = f_1(z_1 + |Q| \cdot x(z_2 - z_1), t_1)$$

and

$$f_2(z_1 + (|Q| \cdot x + m)(z_2 - z_1), t_1 + n(t_2 - t_1))$$
$$= f_2(z_1 + (|Q| \cdot x + m)(z_2 - z_1), t_1)$$
$$+ n(f_2(z_1 + |Q| \cdot x(z_2 - z_1), t_2) - f_2(z_1 + |Q| \cdot x(z_2 - z_1), t_1)),$$

where

$$f_2(z_1 + (|Q| \cdot x + m)(z_2 - z_1), t_1)$$
$$= f_2(z_1 + (|Q| \cdot x + m)(z_2 - z_1), 0) + f_2(z_1 + |Q| \cdot x(z_2 - z_1), t_1)$$
$$- f_2(z_1 + |Q| \cdot x(z_2 - z_1), 0)$$
$$= f_2(z_1 + |Q| \cdot x(z_2 - z_1), t_1) + m(f_2(z_2, 0) - f_2(z_1, 0)).$$

If

$$f_2(z_1 + |Q| \cdot x(z_2 - z_1), t_2) > f_2(z_1 + |Q| \cdot x(z_2 - z_1), t_1),$$

we get

$$f_2(z_1 + (|Q| \cdot x + f_2(z_1 + |Q| \cdot x(z_2 - z_1), t_2)$$
$$- f_2(z_1 + |Q| \cdot x(z_2 - z_1), t_1))(z_2 - z_1), t_1)$$
$$= f_2(z_1 + |Q| \cdot x(z_2 - z_1), t_1 + (f_2(z_2, 0) - f_2(z_1, 0))(t_2 - t_1)).$$

If

$$f_2(z_1 + |Q| \cdot x(z_2 - z_1), t_2) < f_2(z_1 + |Q| \cdot x(z_2 - z_1), t_1),$$

it holds

$$f_2(z_1 + (|Q| \cdot x + f_2(z_1 + |Q| \cdot x(z_2 - z_1), t_1)$$
$$- f_2(z_1 + |Q| \cdot x(z_2 - z_1), t_2))(z_2 - z_1),$$
$$t_1 + (f_2(z_2, 0) - f_2(z_1, 0))(t_2 - t_1))$$
$$= f_2(z_1 + |Q| \cdot x(z_2 - z_1), t_1).$$

Hence, we have a contradiction. □

Example 4.63. Let $n \geq 2$ and L_n be the permutation closed language

$$\{w \in \{a, b, c, d\}^* \mid |w|_b = |w|_c = 0 < |w|_a \leq |w|_d < |w|_a + n - 1$$
$$\vee |w|_d = 0 < |w|_b = |w|_c\}.$$

For $v, w \in L_n$ we have $v \sim_{L_n} w$ if and only if

$$(|v|_b = |w|_b = 0 \wedge |v|_d - |v|_a = |w|_d - |w|_a) \vee |v|_d = |w|_d = 0.$$

Hence, we get $|L_n/{\sim_{L_n}}| = n$. There are $n - 1$ elements $T \in \psi(L_n)/{\sim_{\psi(L_n)}}$ with

$$(T - T) \cap \mathbb{N}^4 = \{(a, 0, 0, a) \mid a \in \mathbb{N}\}$$

and one $U \in \psi(L_n)/{\sim_{\psi(L_n)}}$ with

$$(U - U) \cap \mathbb{N}^4 = \{(a, b, b, 0) \mid a, b \in \mathbb{N}\}.$$

Let $T, U \in \psi(L_n)/{\sim_{\psi(L_n)}}$ with the properties from the last sentence. Then, we get that $\dim(V_T) = 1$, $\dim(V_U) = 2$, $X_T = \{1, 4\}$, and $X_U = \{1, 2, 3\}$. We have $i_T = 1$ and $(0, 0, 0, 1) = (1, 0, 0, 1) - (1, 0, 0, 0)$. This gives us $v_T = (1, 0, 0, -1)$. Moreover, it holds $i_U = 2$ and $(0, 0, 1, 0) = (0, 1, 1, 0) - (0, 1, 0, 0)$. That implies $v_U = (0, 1, -1, 0)$. By Proposition 4.59 the language L_n is accepted by a deterministic counter automaton. On the other hand, the language L_n is not accepted by a λ-free deterministic counter automaton by Proposition 4.62.

Part III

Variants of Jumping Finite Automata

5 Operational State Complexity of Jumping Finite Automata

Operational state complexity investigates problems of the following kind. Consider an automaton model T and a language operation $\circ$ with p operands under which the family of languages accepted by automata of type T is closed. Given automata $A_1, A_2, \ldots, A_p$ of type T, how large can the minimal number of internal states of an automaton of type T accepting the language $\circ(L(A_1), L(A_2), \ldots, L(A_p))$ get, depending on the numbers of states of $A_1, A_2, \ldots, A_p$? These problems were studied exhaustively for finite automata. An overview of the results can be found in a survey on operational state complexity by Yuan Gao *et al.* from 2016 [30]. We give upper bounds for the operational state complexity of jumping finite automata. The family **JFA** is closed under union, intersection, complementation, and inverse homomorphism. Since the operations of permutation closure and union commute, for the operational state complexity of jumping finite automata under union we get the same upper bounds as for finite automata. For intersection, complementation, and inverse homomorphism the situation is much more complicated. The proofs for the closure of **JFA** under these three operations use the fact that semilinear sets are closed under these operations. Thus, our strategy to deduce upper bounds for the operational state complexity of jumping finite automata under these operations is the following. A result by To gives us size estimations for the Parikh-images of the languages accepted by our operand automata [73]. Then, we use our results from Chapter 3 to get upper bounds for the resulting semilinear set when our operation is applied to these Parikh-images. From these bounds for the resulting semilinear set, we conclude upper bounds for the number of states of the resulting jumping finite automaton. To get upper bounds for the operational state complexity of DFAswttf interpreted as jumping finite automata, we use a result by Giovanna J. Lavado *et al.* about the state complexity of converting an NFA to a Parikh-equivalent DFAwttf [49].

Lavado *et al.* also gave an upper bound for the state complexity of the following problem: given two DFAswttf A_1, A_2, construct a DFAwttf A_3 such that for the Parikh-images we have $\psi(L(A_3)) = \psi(L(A_1)) \cap \psi(L(A_2))$ [50]. This is equivalent to giving an upper bound for the operational state complexity of DFAswttf interpreted as jumping finite automata under the operation of intersection. Here, our bound improves the one by Lavado *et al.* Finally, we consider the operation of intersection with regular languages, under which **JFA** is not closed, in the case where the resulting language is in **JFA**. We give a short summary of our upper bounds for the operational state complexity of jumping finite automata: Our bounds for the operations of intersection and intersection with regular languages are polynomial in the numbers of states of the operand automata. Our bound for the operation of inverse homomorphism is polynomial in the number of states of the operand automaton and in the absolute value of the homomorphism. Our bound for complementation is single exponential in the number of states of the operand automaton. All the mentioned bounds also depend on the cardinalities of the input alphabets of the considered jumping finite automata.

We now give some results that we later use to deduce our bounds for the operational state complexity of jumping finite automata. A theorem by To delivers size estimations for the Parikh-image of the language accepted by an NFA:

Theorem 5.1 ([73], Theorem 4.1). *Let A be an n-state NFA with input alphabet of cardinality k. Then, there exists a semilinear representation $R = (I, (\vec{c}_i)_{i \in I}, (P_i)_{i \in I})$ in $\mathbb{N}^k$ such that $S(R) = \psi(L(A))$, $n(R) \leq n$,*

$$K(R) \in O\left(k^{4k+6} n^{k^2+3k+5}\right),$$
$$m(R) \in O\left(k^{4k+6} n^{3k+5}\right),$$

and for all $i \in I$ the set P_i is a linearly independent subset of the k-dimensional Euclidean space.

To's original theorem says that for all $i \in I$ we have $|P_i| \leq k$, but does not state that the sets P_i are linearly independent. By inspecting To's proof, one can however observe that the sets P_i are indeed linearly independent. Notice that all the bounds from Theorem 5.1 are polynomial in n. The next result bounds the number of states of an NFA A such that the Parikh-image of $L(A)$ is a given semilinear set.

Proposition 5.2. *Let Σ be an alphabet and R be a representation through hybrid linear sets in $\mathbb{N}^{|\Sigma|}$. Then, there exists an NFA A with input alphabet Σ and*

$$|\Sigma| \cdot K(R) \cdot (M(R) \cdot m(R) + N(R) \cdot n(R)) + 1$$

states such that $\psi(L(A)) = S(R)$.

Proof. It suffices to construct a λ-NFA with the stated properties because for each λ-NFA there is an equivalent NFA with the same number of states. Let $R = (I, (C_i)_{i \in I}, (P_i)_{i \in I})$. First we fix $i \in I$ and construct a λ-NFA A_i such that $\psi(L(A_i)) = \mathsf{L}(C_i, P_i)$. For every $\vec{v} \in C_i$ there is a λ-NFA $B_{\vec{v}}$ such that the accepted language is a singleton with Parikh-image $\{\vec{v}\}$. If $\vec{0} \in C_i$ the λ-NFA $B_{\vec{0}}$ should consist of an initial and a different accepting state and a λ-edge from the initial state to the accepting state. For $m(R) > 0$ the λ-NFA $B_{\vec{v}}$ can be constructed such that it has not more than $|\Sigma| \cdot m(R) + 1$ states. Next we get a λ-NFA B out of the automata $B_{\vec{v}}$ by gluing together all their initial states to one initial state and gluing together all their accepting states to one accepting state. This gives us $\psi(L(B)) = C_i$. For $m(R) > 0$ the λ-NFA B has at most

$$M(R)\left(|\Sigma| \cdot m(R) + 1 - 2\right) + 2 \leq |\Sigma| \cdot M(R) \cdot m(R) + 1$$

states. For $m(R) = 0$ the λ-NFA B has two states.

Analogously, for every $\vec{w} \in P_i$, we have an NFA $D_{\vec{w}}$ with at most $|\Sigma| \cdot n(R) + 1$ states such that the accepted language is a singleton with Parikh-image $\{\vec{w}\}$. We get an NFA D by gluing together the initial and accepting states of all the $D_{\vec{w}}$ to one state, which is the initial and only accepting state of D. Clearly, we have $\psi(L(D) = \mathsf{L}(\vec{0}, P_i)$. For $n(R) > 0$ the number of states of D is bounded from above by

$$N(R)\left(|\Sigma| \cdot n(R) + 1 - 2\right) + 1 \leq |\Sigma| \cdot N(R) \cdot n(R).$$

For $n(R) = 0$ the NFA D has one state.

If $m(R) > 0$ or $n(R) > 0$ we get the λ-NFA A_i with $\psi(L(A_i)) = \mathsf{L}(C_i, P_i)$ by gluing together the accepting state of B with the initial and accepting state of D to one state, which is the accepting state of A_i. In this case, A_i has at most

$$|\Sigma| \cdot M(R) \cdot m(R) + 1 + |\Sigma| \cdot N(R) \cdot n(R) + 1 - 1$$

states. If $m(R) = n(R) = 0$ we let A_i consist of just one state. Finally, we get the λ-NFA A with $\psi(L(A)) = S(R)$ by gluing together all the initial states of the A_i to one initial state. So A has at most

$$K(R)\left(|\Sigma| \cdot (M(R) \cdot m(R) + N(R) \cdot n(R)) + 1 - 1\right) + 1$$

states. $\qquad\qquad\qquad\qquad\qquad\qquad\qquad\qquad\qquad\qquad\qquad\square$

The following lemma by Lavado *et al.* decomposes the language accepted by a finite automaton into unary and non-unary parts.

Lemma 5.3 ([49], Lemma 2.5). *Let A be an NFA with n states and input alphabet $\Sigma = \{a_1, a_2, \ldots, a_k\}$. Then, for every $i \in \{1, 2, \ldots, k\}$ there exists an NFA A_i with input alphabet $\{a_i\}$ and n states such that*

$$\left(\iota_\Sigma \circ \iota_{\{a_i\} \to \Sigma}\right)^* (L(A_i)) = L(A) \cap \{a_i\}^{*(\Sigma^*, \cdot_\Sigma, \lambda_\Sigma)}.$$

Furthermore, there exists an NFA A_0 with input alphabet Σ and $(k+1)n + 1$ states such that

$$L(A_0) = L(A) \setminus \bigcup_{i=1}^{k} \{a_i\}^{*(\Sigma^*, \cdot_\Sigma, \lambda_\Sigma)}.$$

If A is a DFAwttf, so are the $A_0, A_1, \ldots, A_k$.

For the determinization of finite automata modulo Parikh-equivalence, Lavado *et al.* showed:

Theorem 5.4 ([49], Theorems 3.3 and 3.4). *Let A be an NFA with n states and an input alphabet of cardinality k. Then, there is a Parikh-equivalent DFAwttf with $e^{\Theta\left(\sqrt{n \cdot \ln(n)}\right)}$ states. If every word in $L(A)$ contains at least two different symbols, there is a DFAwttf with*

$$O\left(k^{k^3/2 + k^2 + 2k + 5} n^{3k^3 + 6k^2}\right)$$

states which is Parikh-equivalent to A.

Notice that the bound $e^{\Theta\left(\sqrt{n \cdot \ln(n)}\right)}$ is tight already in the unary case as shown by Marek Chrobak [20]. The bound given in Theorem 5.4 for the case that every word in $L(A)$ contains at least two different symbols is poynomial in n. Hence, in this case determinization modulo Parikh-equivalence can be done much more efficient than in the general case if the input alphabet is fixed. Theorem 5.4 can also be read as a result about converting an NFA A to a DFAwttf which is equivalent to A when both automata are interpreted as jumping finite automata.

5.1 Intersection

For the state complexity of the intersection of the languages accepted by two n state NFAs over an input alphabet of cardinality k interpreted as jumping finite automata we get an upper bound of order $(kn)^{O\left(k^2\right)}$.

Corollary 5.5. *Let A and B be two NFAs over the same input alphabet of cardinality k and let n be the maximum of the numbers of states of A and B. Then, there is an NFA C with*

$$O\left(k^{9k^2/2 + 43k/2 + 21} n^{6k^2 + 15k + 15}\right)$$

states such that $L_J(C) = L_J(A) \cap L_J(B)$.

Proof. Due to Theorem 5.1 there exist two semilinear representations R and T in $\mathbb{N}^k$ such that it holds $S(R) = \psi(L(A))$, $S(T) = \psi(L(B))$,

$\max(N(R), N(T)) \leq k$, $\max(n(R), n(T)) \leq n$,

$$K(R), K(T) \in O\left(k^{4k+6} n^{k^2+3k+5}\right),$$
$$m(R), m(T) \in O\left(k^{4k+6} n^{3k+5}\right).$$

By Theorem 3.4 there is a representation U through hybrid linear sets in $\mathbb{N}^k$ with $S(U) = \psi(L(A)) \cap \psi(L(B))$,

$$K(U) \in O\left(\left(k^{4k+6} n^{k^2+3k+5}\right)^2\right) = O\left(k^{8k+12} n^{2k^2+6k+10}\right),$$
$$m(U), n(U) \in O\left(k^{k/2} k^2 n^k k^{4k+6} n^{3k+5}\right) = O\left(k^{9k/2+8} n^{4k+5}\right).$$

There is a constant $d \in \mathbb{N}$ such that

$$\max(m(U), n(U)) + 1 \leq dk^{9k/2+8} n^{4k+5}.$$

We get

$$\max(M(U), N(U)) \leq \left(dk^{9k/2+8} n^{4k+5}\right)^k$$
$$= d^k k^{9k^2/2+8k} n^{4k^2+5k} \in O\left(k^{9k^2/2+9k} n^{4k^2+5k}\right).$$

Now Proposition 5.2 tells us that there is an NFA C with

$$O\left(k\left(k^{8k+12} n^{2k^2+6k+10}\right)\left(k^{9k^2/2+9k} n^{4k^2+5k}\right)\left(k^{9k/2+8} n^{4k+5}\right)\right)$$
$$= O\left(\left(k^{8k+13} n^{2k^2+6k+10}\right)\left(k^{9k^2/2+27k/2+8} n^{4k^2+9k+5}\right)\right)$$
$$= O\left(k^{9k^2/2+43k/2+21} n^{6k^2+15k+15}\right)$$

states such that $\psi(L(C)) = \psi(L(A)) \cap \psi(L(B))$. It follows

$$L_J(C) = \psi^{-1}(\psi(L(C)))$$
$$= \psi^{-1}(\psi(L(A))) \cap \psi^{-1}(\psi(L(B))) = L_J(A) \cap L_J(B).$$

$\square$

If we consider DFAswttf instead of NFAs, we get an upper bound of order $(kn)^{O(k^5)}$.

Corollary 5.6. *Let A and B be two DFAswttf over the same input alphabet of cardinality k and let n be the maximum of the numbers of states of A and B. Then, there is a DFAwttf C with*

$$O\left(k^{27k^5/2+183k^4/2+196k^3}n^{18k^5+81k^4+135k^3+90k^2+2}\right)$$

states such that $L_J(C) = L_J(A) \cap L_J(B)$.

Proof. Let $\Sigma = \{a_1, a_2, \ldots, a_k\}$ be the common input alphabet of A and B and $i \in \{1, 2, \ldots, k\}$. By Lemma 5.3 there exist n-state DFAswttf A_i and B_i with input alphabet $\{a_i\}$ such that

$$\left(\iota_\Sigma \circ \iota_{\{a_i\} \to \Sigma}\right)^* \left(L_J(A_i)\right) = L_J(A) \cap \{a_i\}^{*(\Sigma^*, \Sigma, \lambda_\Sigma)}$$

and

$$\left(\iota_\Sigma \circ \iota_{\{a_i\} \to \Sigma}\right)^* \left(L_J(B_i)\right) = L_J(B) \cap \{a_i\}^{*(\Sigma^*, \Sigma, \lambda_\Sigma)}.$$

Since $L_J(A_i)$ and $L_J(B_i)$ are unary, by the cross-product construction we obtain a DFAwttf C_i with input alphabet $\{a_i\}$ and n^2 states such that $L_J(C_i) = L_J(A_i) \cap L_J(B_i)$. Without loss of generality we may assume that the state sets of the automata C_i are pairwise disjoint. We now construct a DFAwttf C with $kn^2 + 2$ states and input alphabet Σ such that

$$L_J(C) = \bigcup_{i=1}^{k} \left(\iota_\Sigma \circ \iota_{\{a_i\} \to \Sigma}\right)^* \left(L_J(C_i)\right) = L_J(A) \cap L_J(B) \cap \bigcup_{i=1}^{k} \{a_i\}^{*(\Sigma^*, \Sigma, \lambda_\Sigma)}.$$

The state set of C consists of the union of the state sets of the C_i (That are k automata, each with n^2 states.), a new start state s, and a new sink state d. The final states of C are the final states of the C_i. In addition, the start state s is a final state if and only if there exists an $i \in \{1, 2, \ldots, k\}$ with $\lambda \in L_J(C_i)$. All the transitions of the C_i also exist in C. Additionally, for every $i \in \{1, 2, \ldots, k\}$ there is an a_i-edge from s to the unique state of C_i that is entered when reading a_i in the start state of C_i. For $i, j \in \{1, 2, \ldots, k\}$ with $i \neq j$ there is an a_j-edge from each state of C_i to d. Finally, for every $i \in \{1, 2, \ldots, k\}$ there is an a_i-edge from d to d.

Because of Corollary 5.5 there exists an NFA with

$$O\left(k^{9k^2/2+43k/2+21}n^{6k^2+15k+15}\right)$$

states, which, interpreted as a jumping finite automaton, accepts $L_J(A) \cap L_J(B)$. So, by Lemma 5.3 there exists an NFA D with

$$O\left(k^{9k^2/2+43k/2+22}n^{6k^2+15k+15}\right)$$

states such that $L_J(D) = (L_J(A) \cap L_J(B)) \setminus L_J(C)$. Now, from Theorem 5.4 it follows that there exists a constant $c \in \mathbb{N}$ and a DFAwttf D_0 with

$$c^{3k^3+6k^2}k^{k^3/2+k^2+2k+5+(9k^2/2+43k/2+22)(3k^3+6k^2)}n^{(6k^2+15k+15)(3k^3+6k^2)}$$

$$= c^{3k^3+6k^2}k^{27k^5/2+183k^4/2+391k^3/2+133k^2+2k+5}n^{18k^5+81k^4+135k^3+90k^2}$$

states such that $L_J(D_0) = (L_J(A) \cap L_J(B)) \setminus L_J(C)$. Since the language operation of union commutes with the permutation closure, there exists a DFAwttf with

$$c^{3k^3+6k^2}k^{27k^5/2+183k^4/2+391k^3/2+133k^2+2k+6}n^{18k^5+81k^4+135k^3+90k^2+2}$$

$$\in O\left(k^{27k^5/2+183k^4/2+196k^3}n^{18k^5+81k^4+135k^3+90k^2+2}\right)$$

states, which, interpreted as a jumping finite automaton, accepts $L_J(C) \cup L_J(D_0) = L_J(A) \cap L_J(B)$. For the latter size estimation we used that, if k is large enough, it holds $c^{3k^3+6k^2}k^{133k^2+2k+6} \leq k^{k^3/2}$. $\qquad\square$

Corollary 5.6 improves a result by Lavado *et al.* which tells us that for each $k > 0$ there is a polynomial p_k such that in the situation of Corollary 5.6 the value $p_k(n)$ is an upper bound for the number of states of C [50, Theorem 10].

5.2 Complementation

For the state complexity of the complementation of the language accepted by a jumping finite automaton with n states and an input alphabet of cardinality k we get an upper bound of order $2^{n^{O(k^2 \cdot \log(k))}}$.

Corollary 5.7. *Let A be an n-state NFA over an input alphabet Σ of cardinality k. Then, there is a DFAwttf B with*

$$2^{k^{(4k+6)\log(3k+1)+O(1)}n^{(k^2+3k+6)\log(3k+1)}}$$

states such that $L_J(B) = \Sigma^* \setminus L_J(A)$.

Proof. The corollary is clearly true for $n = 1$ or $k = 1$, so let $n > 1$ and $k > 1$. Assume that $L(A)$ is finite. Then, for every $w \in L_J(A)$ it holds $|w| \leq n - 1$. So, there is a DFAwttf B with

$$\sum_{i=0}^{n-1} k^i + 1 \leq k^n = 2^{\log(k)\cdot n}$$

states such that $L(B) = \Sigma^* \setminus L_J(A)$. It follows $L_J(B) = \mathsf{perm}(L(B)) = \Sigma^* \setminus L_J(A)$. Hence, the corollary is true if $L(A)$ is finite. Thus, let $L(A)$ be infinite in the following. By Theorem 5.1 there is a semilinear representation $R = (I, (\vec{c}_i)_{i\in I}, (P_i)_{i\in I})$ in $\mathbb{N}^k$ such that $S(R) = \psi(L(A))$, $n(R) \leq n$,

$$K(R) \in O\left(k^{4k+6}n^{k^2+3k+5}\right),$$
$$m(R) \in O\left(k^{4k+6}n^{3k+5}\right),$$

and for all $i \in I$ the set P_i is a linearly independent subset of the k-dimensional Euclidean space. Lemma 3.10 gives us a representation R' through hybrid linear sets in $\mathbb{N}^k$ with $S(R') = \mathbb{N}^k \setminus \psi(L(A))$,

$$K(R') \in 2^{O\left(k^{4k+7}n^{k^2+3k+5}\right)},$$
$$n(R') \in (4kn)^{(k+2)\cdot\left(O\left(k^{4k+6}n^{k^2+3k+5}\right)\right)^{\log(3k+1)}},$$
$$m(R') \in (4kn)^{(k+2)\cdot\left(O\left(k^{4k+6}n^{k^2+3k+5}\right)\right)^{\log(3k+1)}} \cdot O\left(k^{4k+6}n^{3k+5}\right).$$

For every $c \in \mathbb{N}_{>0}$ it holds

$$\log(4kn)(k+2)c^{\log(3k+1)} \leq \log(4kn)(k+2)c^{\log(4k)}$$
$$= c^2(\log(4k) + \log(n))(k+2)k^{\log(c)}.$$

This gives us

$$n(R'), m(R') \in 2^{k^{(4k+6)\log(3k+1)+O(1)}n^{(k^2+3k+6)\log(3k+1)}},$$

which implies

$$N(R'), M(R') \in 2^{k^{(4k+6)\log(3k+1)+O(1)}n^{(k^2+3k+6)\log(3k+1)}}.$$

By Proposition 5.2 there is an NFA B with input alphabet Σ and

$$2^{k^{(4k+6)\log(3k+1)+O(1)}n^{(k^2+3k+6)\log(3k+1)}}$$

states such that $\psi(L(B)) = S(R')$. It follows

$$L_J(B) = \psi^{-1}(\psi(L(B))) = \psi^{-1}\left(\mathbb{N}^k \setminus \psi(L(A))\right)$$
$$= \Sigma^* \setminus \psi^{-1}(\psi(L(A))) = \Sigma^* \setminus L_J(A).$$

Let $\Sigma = \{a_1, a_2, \ldots, a_k\}$ and $i \in \{1, 2, \ldots, k\}$. Lemma 5.3 tells us that there exists an NFA A_i with input alphabet $\{a_i\}$ and n states such that

$$\left(\iota_\Sigma \circ \iota_{\{a_i\} \to \Sigma}\right)^* (L(A_i)) = L(A) \cap \{a_i\}^{*(\Sigma^*, \cdot_\Sigma, \lambda_\Sigma)}.$$

The NFA A_i can be determinized into a DFAwttf with at most 2^n states that is further converted into a DFAwttf C_i with the same number of states and $L(C_i) = \{a_i\}^* \setminus L(A_i)$. By using a copy of each C_i, a new initial state that selects the appropriate C_i dependent on the first input symbol, and a new sink state that is entered whenever a word contains two different symbols, a $(k2^n + 2)$-state DFAwttf C over the input alphabet Σ is constructed such that

$$L_J(C) = \bigcup_{i=1}^{k} \left(\iota_\Sigma \circ \iota_{\{a_i\} \to \Sigma}\right)^* (L(C_i))$$
$$= \bigcup_{i=1}^{k} \left(\{a_i\}^{*(\Sigma^*, \cdot_\Sigma, \lambda_\Sigma)} \setminus \left(\iota_\Sigma \circ \iota_{\{a_i\} \to \Sigma}\right)^* (L(A_i))\right)$$
$$= \left(\bigcup_{i=1}^{k} \{a_i\}^{*(\Sigma^*, \cdot_\Sigma, \lambda_\Sigma)}\right) \setminus L(A)$$
$$= \left(\bigcup_{i=1}^{k} \{a_i\}^{*(\Sigma^*, \cdot_\Sigma, \lambda_\Sigma)}\right) \setminus L_J(A)$$
$$= (\Sigma^* \setminus L_J(A)) \cap \bigcup_{i=1}^{k} \{a_i\}^{*(\Sigma^*, \cdot_\Sigma, \lambda_\Sigma)}.$$

By Lemma 5.3 there is an NFA B_0 with

$$2^{k^{(4k+6)\log(3k+1)+O(1)}n^{(k^2+3k+6)\log(3k+1)}}$$

states such that

$$L_J(B_0) = L_J(B) \setminus \bigcup_{i=1}^{k} \{a_i\}^{*(\Sigma^*, \cdot_\Sigma, \lambda_\Sigma)} = (\Sigma^* \setminus L_J(A)) \setminus \bigcup_{i=1}^{k} \{a_i\}^{*(\Sigma^*, \cdot_\Sigma, \lambda_\Sigma)}.$$

From Theorem 5.4 it follows that there exists a DFAwttf B_0' with

$$O\left(k^{k^3/2+k^2+2k+5}\left(2^{k^{(4k+6)\log(3k+1)+O(1)}n^{(k^2+3k+6)\log(3k+1)}}\right)^{3k^3+6k^2}\right)$$

$$= O\left(2^{\log(k)(k^3/2+k^2+2k+5)}2^{k^{(4k+6)\log(3k+1)+O(1)}n^{(k^2+3k+6)\log(3k+1)}}\right)$$

$$= 2^{k^{(4k+6)\log(3k+1)+O(1)}n^{(k^2+3k+6)\log(3k+1)}}$$

states such that $L_J(B_0') = L_J(B_0)$. Because the language operation of union commutes with the permutation closure, there is a DFAwttf D with

$$(k2^n+2)2^{k^{(4k+6)\log(3k+1)+O(1)}n^{(k^2+3k+6)\log(3k+1)}} = 2^{k^{(4k+6)\log(3k+1)+O(1)}n^{(k^2+3k+6)\log(3k+1)}}$$

states and $L_J(D) = L_J(C) \cup L_J(B_0') = \Sigma^* \setminus L_J(A)$. $\qquad\square$

5.3 Inverse Homomorphism

For the state complexity of the preimage of the language accepted by an n state NFA over an input alphabet Σ_2 of cardinality k_2 interpreted as a jumping finite automaton under a monoid homomorphism $h : \Sigma_1^* \to \Sigma_2^*$, where Σ_1 is an alphabet of cardinality k_1, we get an upper bound of order $(\max{(k_1, k_2, |h|, n)})^{k_2^2+O(k_1 k_2)}$.

Corollary 5.8. *Let Σ_1 and Σ_2 be alphabets of cardinalities k_1 and k_2, $h : \Sigma_1^* \to \Sigma_2^*$ be a monoid homomorphism, and A be an n-state NFA over Σ_2. Then, there is an NFA B over Σ_1 with*

$$(O(k_1))^{k_1+2}k_2^{9k_1k_2/2+7k_1+17k_2/2+13}n^{3k_1k_2+k_2^2+5k_1+6k_2+10}(\max{(n, |h|)})^{k_1k_2+k_2}$$

states such that $L_J(B) = h^{-1}(L_J(A))$.

Proof. Theorem 5.1 tells us that there exists a semilinear representation $R = (I, (\vec{c_i})_{i \in I}, (P_i)_{i \in I})$ in $\mathbb{N}^{k_2}$ such that it holds $S(R) = \psi_{\Sigma_2}(L(A))$, $n(R) \leq n$, $N(R) \leq k_2$,

$$K(R) \in O\left(k_2^{4k_2+6} n^{k_2^2+3k_2+5}\right),$$

and

$$m(R) \in O\left(k_2^{4k_2+6} n^{3k_2+5}\right).$$

Let $\Sigma_1 = \{a_1, a_2, \ldots, a_{k_1}\}$ and $H : \mathbb{N}^{k_1} \to \mathbb{N}^{k_2}$ be the monoid homomorphism such that for all $j \in \{1, 2, \ldots, k_1\}$ it holds $H\left(\overrightarrow{e_{k_1,j}}\right) = \psi_{\Sigma_2}(h(a_j))$. From $H \circ \psi_{\Sigma_1} = \psi_{\Sigma_2} \circ h$ it follows

$$\psi_{\Sigma_1}\left(h^{-1}\left(L_J(A)\right)\right) = H^{-1}\left(\psi_{\Sigma_2}\left(L_J(A)\right)\right) = H^{-1}(S(R)).$$

Now, Proposition 3.16 gives us a representation R' through hybrid linear sets in $\mathbb{N}^{k_1}$ with $S(R') = H^{-1}(S(R))$,

$$K(R') \in O\left(k_2^{4k_2+6} n^{k_2^2+3k_2+5}\right),$$

$$n(R') \leq (k_1 + k_2 + 1)\left(\sqrt{k_2} \cdot \max\left(n, |h|\right)\right)^{k_2},$$

and

$$m(R') \in O\left(k_1 k_2^{9k_2/2+7} n^{3k_2+5}\left(\max\left(n, |h|\right)\right)^{k_2}\right).$$

So, there is a $c \in \mathbb{N}$ with

$$\max(n(R'), m(R')) + 1 \leq c k_1 k_2^{9k_2/2+7} n^{3k_2+5}\left(\max\left(n, |h|\right)\right)^{k_2}.$$

It follows

$$\max(N(R'), M(R')) \leq \left(c k_1 k_2^{9k_2/2+7} n^{3k_2+5}\left(\max\left(n, |h|\right)\right)^{k_2}\right)^{k_1}$$

$$\in (O(k_1))^{k_1} k_2^{9k_1 k_2/2+7k_1} n^{3k_1 k_2+5k_1}\left(\max\left(n, |h|\right)\right)^{k_1 k_2}.$$

By Proposition 5.2 there exists an NFA B with input alphabet Σ_1 and

$$k_1 \cdot k_2^{4k_2+6} n^{k_2^2+3k_2+5} \cdot k_1 k_2^{9k_2/2+7} n^{3k_2+5}\left(\max\left(n, |h|\right)\right)^{k_2}$$

$$\cdot (O(k_1))^{k_1} k_2^{9k_1 k_2/2+7k_1} n^{3k_1 k_2+5k_1}\left(\max\left(n, |h|\right)\right)^{k_1 k_2}$$

$$= (O(k_1))^{k_1+2} k_2^{9k_1 k_2/2+7k_1+17k_2/2+13} n^{3k_1 k_2+k_2^2+5k_1+6k_2+10}\left(\max\left(n, |h|\right)\right)^{k_1 k_2+k_2}$$

states such that $\psi_{\Sigma_1}(L(B)) = S(R')$. We get

$$L_J(B) = \psi_{\Sigma_1}^{-1}(\psi_{\Sigma_1}(L(B))) = \psi_{\Sigma_1}^{-1}\left(\psi_{\Sigma_1}\left(h^{-1}(L_J(A))\right)\right) = h^{-1}(L_J(A)).$$

$\square$

5.4 Intersection with Regular Languages

The language family **JFA** is not closed under intersection with regular languages because for every alphabet Σ and every regular language $R \subset \Sigma^*$ which is not permutation closed we get the counterexample $\Sigma^* \cap R = R$. However, we show that it is decidable if the intersection of a language from **JFA** and a regular language is again in **JFA**. In this case we give upper bounds for the operational state complexity. We have the following criterion when the intersection of a language from **JFA** and a regular language is again in **JFA**.

Lemma 5.9. *Let A be an NFA with input alphabet Σ and $R \subseteq \Sigma^*$ be a regular language. Then, the following statements are equivalent:*

1. *$L_J(A) \cap R \in$ **JFA**.*

2. *$L_J(A) \cap R = L_J(A) \cap \mathsf{perm}(R)$*

3. *$L_J(A) \cap \mathsf{perm}(R) \cap \mathsf{perm}(\Sigma^* \setminus R) = \emptyset$*

Proof. We start proving "1. $\Rightarrow$ 3.:" Assume that $L_J(A) \cap R \in$ **JFA** and consider a word $x \in L_J(A) \cap \mathsf{perm}(R)$. Then there is an $y \in \mathsf{perm}(\{x\}) \cap R$. We have

$$\mathsf{perm}(x) = \mathsf{perm}(y) \subseteq \mathsf{perm}(\mathsf{perm}(\{x\}) \cap R)$$
$$\subseteq \mathsf{perm}(L_J(A) \cap R) = L_J(A) \cap R \subseteq R.$$

It follows $x \notin \mathsf{perm}(\Sigma^* \setminus R)$.

Next we show "3. $\Rightarrow$ 2.:" From the third condition it follows that

$$L_J(A) \cap \mathsf{perm}(R) \subseteq \Sigma^* \setminus \mathsf{perm}(\Sigma^* \setminus R) \subseteq \Sigma^* \setminus (\Sigma^* \setminus R) = R.$$

This implies the second condition.

Finally consider "2. $\Rightarrow$ 1.:" We have

$$\psi_\Sigma(L_J(A) \cap R) = \psi_\Sigma\left(\psi_\Sigma^{-1}\left(\psi_\Sigma(L(A))\right) \cap R\right) = \psi_\Sigma(L(A)) \cap \psi_\Sigma(R),$$

which is semilinear as the intersection of two semilinear sets. On the other hand, the language $L_J(A) \cap \mathsf{perm}(R)$ is permutation closed as the intersection of two permutation closed languages. So, the second condition implies the first one. $\qquad\square$

From Lemma 5.9 it follows that it is decidable if the intersection of a language from **JFA** and a regular language is again in **JFA**:

Proposition 5.10. *For two NFAs A and B over a common input alphabet it is decidable whether $L_J(A) \cap L(B) \in$ **JFA**.*

Proof. Let Σ be the common input alphabet of A and B. The powerset construction gives us a DFAwttf which is equivalent to B. After exchanging accepting and non-accepting states, we get a DFAwttf C with $L(C) = \Sigma^* \setminus L(B)$. By Lemma 5.9 it holds $L_J(A) \cap L(B) \in$ **JFA** if and only if

$$\emptyset = \bigcap_{X \in \{A,B,C\}} \psi_\Sigma^{-1}(\psi_\Sigma(L(X))) = \psi_\Sigma^{-1}\left(\bigcap_{X \in \{A,B,C\}} \psi_\Sigma(L(X))\right).$$

So, we have $L_J(A) \cap L(B) \in$ **JFA** if and only if $\psi_\Sigma(L(A)) \cap \psi_\Sigma(L(B)) \cap \psi_\Sigma(L(C))$ is empty. For a finite automaton we can construct an equivalent regular expression, from which we can easily construct a semilinear representation of the Parikh-image. Ginsburg and Spanier showed that the family of semilinear sets is effectively closed under the operation of intersection [32]. Hence, we can construct a semilinear representation of $\psi_\Sigma(L(A)) \cap \psi_\Sigma(L(B)) \cap \psi_\Sigma(L(C))$. Clearly, we can decide for a given semilinear representation whether it represents the empty set. $\qquad\square$

For the state complexity of the intersection of the languages accepted by an n state NFA over an input alphabet of cardinality k interpreted as a jumping finite automaton and an n state NFA over the same input alphabet we get the same upper bound as in Corollary 5.5 if this intersection is again accepted by an NFA interpreted as a jumping finite automaton:

Corollary 5.11. *Let A and B be two NFAs over the same input alphabet of cardinality k such that $L_J(A) \cap L(B) \in \mathbf{JFA}$ and let n be the maximum of the numbers of states of A and B. Then, there is an NFA C with*

$$O\left(k^{9k^2/2+43k/2+21}n^{6k^2+15k+15}\right)$$

states such that $L_J(C) = L_J(A) \cap L(B)$.

Proof. By Lemma 5.9 it holds $L_J(A) \cap L(B) = L_J(A) \cap L_J(B)$. Now the result follows from Corollary 5.5. □

If we replace NFAs by DFAswttf, we get the same upper bound as in Corollary 5.6:

Corollary 5.12. *Let A and B be two DFAswttf over the same input alphabet of cardinality k such that $L_J(A) \cap L(B) \in \mathbf{JFA}$ and let n be the maximum of the numbers of states of A and B. Then, there is a DFAwttf C with*

$$O\left(k^{27k^5/2+183k^4/2+196k^3}n^{18k^5+81k^4+135k^3+90k^2+2}\right)$$

states such that $L_J(C) = L_J(A) \cap L(B)$.

Proof. By Lemma 5.9 it holds $L_J(A) \cap L(B) = L_J(A) \cap L_J(B)$. Now the result follows from Corollary 5.6. □

6 Properties of Right One-Way Jumping Finite Automata

In the paper in which Chigahara *et al.* introduced right one-way jumping finite automata they gave some results concerning the computational capacity of these devices [14]: The family of languages accepted by such automata, **ROWJ**, is properly sandwiched between the regular and the context-sensitive languages. Furthermore, **ROWJ** is incomparable to the context-free and to the deterministic context-free languages. It was stated as an open problem by Chigahara *et al.* if **JFA** is properly included in **ROWJ** or if these two families are incomparable [14].

We give more results about the computational capacity of right one-way jumping finite automata. Notice that Example 4.2 shows that there is no finitely describable automaton model for the family of all (strongly) semirecognizable languages. However, for permutation closed languages our characterization result, which is an analogy of Theorem 2.19, reads as follows. For all $n \geq 0$ a permutation closed language is in $\mathbf{ROWJ_n}$ if and only if it is n-semirecognizable. Thus, the permutation closed languages accepted by right one-way jumping finite automata are exactly the permutation closed semirecognizable languages. Our characterization of the family **pROWJ** implies that **ROWJ** and **JFA** are incomparable and that **pROWJ** is incomparable to **pCF** and to **pDCF**.

Chigahara *et al.* showed that **ROWJ** is not closed under intersection, intersection with regular languages, reversal, Kleene star, Kleene plus, concatenation, concatenation with a regular language from the right, and substitution [14]. They also conjectured that **ROWJ** is not closed under union. We prove that **ROWJ** is not closed under union, union with regular languages, complementation, concatenation with a regular language from the left, λ-free homomorphism, inverse homomorphism, and permutation closure. However, we show that **ROWJ** is closed under concatenation with a prefix-free regular language from the left. We also give closure properties of p**ROWJ**. They follow from our characterization of p**ROWJ** as the family of all permutation closed semirecognizable languages. Concerning the concatenation of languages we show the following. For an alphabet Σ, a language $L \subseteq \Sigma$, and a word $w \in \Sigma^* \setminus \{\lambda\}$ it holds: The language $\{w\}L$ is in **ROWJ** if and only if L is in **ROWJ**. The language $L\{w\}$ is in **ROWJ** if and only if L is regular. For alphabets Σ, Σ_1, and Σ_2 with $\Sigma_1 \cap \Sigma_2 = \emptyset$ and $\Sigma = \Sigma_1 \cup \Sigma_2$ and languages $L_1 \subseteq \Sigma_1^{*(\Sigma^*, \Sigma, \lambda_\Sigma)}$ and $L_2 \subseteq \Sigma_2^{*(\Sigma^*, \Sigma, \lambda_\Sigma)}$ with $L_1 \neq \emptyset \neq L_2 \neq \{\lambda\}$ we get: if $L_1L_2 \in$ **ROWJ**, then $L_1 \in$ **REG** and $L_2 \in$ **ROWJ**.

A basic property of right one-way jumping finite automata that is used later on is shown now:

Lemma 6.1. *Let $A = (Q, \Sigma, \delta, \{s_0\}, F)$ be a DFA, $v, w \in \Sigma^*$, $p, q \in Q$, and $n \geq 0$ with $(p, v) \circlearrowright^n (q, w)$. Then, there is an $x \in \Sigma^*$ with $|x| \leq n$ such that xw is a permutation of v and $(p, x) \vdash^* (q, \lambda)$.*

Proof. We prove this by induction on n. If $n = 0$, we have $(p, v) = (q, w)$ and just set $x = \lambda$. Now, assume $n > 0$ and that the lemma is true for the relation $\circlearrowright^{n-1}$. We get $r \in Q$, $a \in \Sigma$, and $y \in \Sigma^*$ such that $(p, v) \circlearrowright^{n-1} (r, ay) \circlearrowright (q, w)$. By the induction hypothesis, there is an $x' \in \Sigma^*$ with $|x'| \leq n - 1$ such that $x'ay$ is a permutation of v and $(p, x') \vdash^* (r, \lambda)$. If $\delta(r, a)$ is undefined, we have $r = q$, $w = ya$, and $x'w$ is a permutation of v. Otherwise, it holds $\delta(r, a) = q$, $y = w$, and $(p, x'a) \vdash^* (r, a) \vdash (q, \lambda)$. $\square$

For the complexity of **ROWJ** we get:

Proposition 6.2. *It holds* **ROWJ** $\subset$ **DST**(n, n^2).

Proof. The computation of a DFA A interpreted as a right one-way jumping finite automaton can be simulated by a deterministic multi-tape Turing machine M which has a read-only input tape in the following way: The letters of the input word are copied to a working tape at the beginning and remain in their position during the whole computation. The head of this tape starts at the leftmost input letter and moves from the left to the right simulating the behaviour of A. When the head arrives at the end of the word, it jumps to the leftmost letter and continues the computation from this position. Each letter that is read by A is marked by M and is no longer considered. The computation of M ends when no letters where marked in one sweep of M from the left to the right. Then, the input is accepted by M if and only if all input letters are marked and A is in an accepting state. Clearly, M simulates A and works in linear space and quadratic time. This shows the inclusion in our proposition. The properness of the inclusion is shown by the non-semilinear language $\{\, a^{m^2} \mid m \geq 0 \,\}$. Since for every $m \geq 0$ it holds $(m+1)^2 = m^2 + 2m + 1$, we have $\{\, a^{m^2} \mid m \geq 0 \,\} \in \mathbf{DST}(\log(n), n)$. $\qquad\square$

Since each language in **ROWJ** contains a Parikh-equivalent regular sublanguage, we get:

Proposition 6.3. *A letter bounded language is in* **ROWJ** *if and only if it is regular.* $\qquad\square$

6.1 Permutation Closed Languages, ROWJFAs, and Semirecognizability

Our characterization of $\mathbf{pROWJ_n}$ shows that a permutation closed language is accepted by a right one-way jumping finite automaton if and only if it is semirecognizable:

Theorem 6.4. *For each* $n \geq 0$ *the family* $\mathbf{pROWJ_n}$ *consists of all permutation closed n-semirecognizable languages.*

Proof. Let $A = (Q, \Sigma, \delta, \{s_0\}, F)$ be a DFA with $L_R(A) = \mathsf{perm}(L_R(A))$. Consider $v, w \in L_R(A)$ and $f \in F$ with $(s_0, v) \circlearrowright^* (f, \lambda)$ and $(s_0, w) \circlearrowright^* (f, \lambda)$. Lemma 6.1 shows that there are permutations v' and w' of v

and w with $(s_0, v') \vdash^* (f, \lambda)$ and $(s_0, w') \vdash^* (f, \lambda)$. Let $x, y \in \Sigma^*$. We get $(s_0, v'xy) \circlearrowright^* (f, xy)$ and $(s_0, w'xy) \circlearrowright^* (f, xy)$. Because $L_R(A)$ is closed under permutation, we have the equivalence

$$xvy \in L_R(A) \Leftrightarrow v'xy \in L_R(A) \Leftrightarrow w'xy \in L_R(A) \Leftrightarrow xwy \in L_R(A).$$

It follows $v \sim_{L_R(A)} w$. From

$$L_R(A) = \bigcup_{f \in F} \left\{ w \in \Sigma^* \mid (s_0, w) \circlearrowright^* (f, \lambda) \right\}$$

we get $\left| L_R(A)/\sim_{L_R(A)} \right| \leq |F|$, which means that the language $L_R(A)$ is $|F|$-semirecognizable.

Let now $k \geq 1$, $\Sigma = \{a_1, a_2, \ldots, a_k\}$ be an alphabet, and $L \subseteq \Sigma^*$ be permutation closed and semirecognizable. Set $L_\lambda = L \cup \{\lambda\}$ and define the mapping $S : L_\lambda/\sim_L \to 2^{\mathbb{N}^k}$ through

$$[w] \mapsto \left\{ \vec{x} \in \mathbb{N}^k \setminus \vec{0} \mid \psi^{-1}(\{\psi(w) + \vec{x}\}) \subseteq L \right\}.$$

The definition of $\sim_L$ and the fact that L is closed under permutation make the map S well-defined. For each $[w] \in L_\lambda/\sim_L$ let $M([w])$ be the set of minimal elements of $S([w])$. So, for every $[w] \in L_\lambda/\sim_L$ and $\vec{x} \in S([w])$, there is an $\vec{x_0} \in M([w])$ such that $\vec{x_0} \leq \vec{x}$. For each $i \in \{1, 2, \ldots, k\}$ let $\pi_i : \mathbb{N}^k \to \mathbb{N}$ be the canonical projection on the ith factor and set

$$m_i = \max \left(\bigcup_{[w] \in L_\lambda/\sim_L} \left\{ \pi_i(\vec{x}) \mid \vec{x} \in M([w]) \right\} \right).$$

Let

$$Q = \left\{ q_{[wv]_{\sim_L}} \mid w \in L_\lambda,\, v \in \Sigma^*,\, \forall i \in \{1, 2, \ldots, k\} : |v|_{a_i} \leq m_i \right\}$$

be a set of states. The finiteness of $L_\lambda/\sim_L$ implies that Q is also finite. Set

$$F = \left\{ q_{[w]_{\sim_L}} \mid w \in L \right\} \subseteq Q.$$

Define the partial mapping $\delta : Q \times \Sigma \to Q$ as follows. For $a \in \Sigma$ and $y \in \Sigma^*$ such that $q_{[y]_{\sim_L}} \in Q$ let

$$\delta \left(q_{[y]_{\sim_L}}, a \right) = q_{[ya]_{\sim_L}}$$

if $q_{[ya]_{\sim L}} \in Q$ and $\delta\left(q_{[y]_{\sim L}}, a\right)$ be undefined otherwise. The DFA

$$A = \left(Q, \Sigma, \delta, \left\{q_{[\lambda]_{\sim L}}\right\}, F\right)$$

is shown to fulfill $L_R(A) = L$.

First, let $y \in L_R(A)$. Then, there exists $w \in L$ with $\left(q_{[\lambda]_{\sim L}}, y\right) \circlearrowright^*$ $\left(q_{[w]_{\sim L}}, \lambda\right)$. From Lemma 6.1 it follows that there is a permutation y' of y with $\left(q_{[\lambda]_{\sim L}}, y'\right) \vdash^* \left(q_{[w]_{\sim L}}, \lambda\right)$. The definition of δ tells us $y' \sim_L w$. We get $y' \in L$ and also $y \in L$ because L is closed under permutation. That shows $L_R(A) \subseteq L$.

Now, let $y \in \Sigma^* \setminus L_R(A)$. There are two possibilities:

1. There is a $w \in \Sigma^* \setminus L$ with $q_{[w]_{\sim L}} \in Q$ such that $\left(q_{[\lambda]_{\sim L}}, y\right) \circlearrowright^*$ $\left(q_{[w]_{\sim L}}, \lambda\right)$. Then, there is a permutation y' of y with $\left(q_{[\lambda]_{\sim L}}, y'\right) \vdash^*$ $\left(q_{[w]_{\sim L}}, \lambda\right)$. We get $y' \sim_L w$. It follows $y' \notin L$, which gives us $y \notin L$.

2. There is a $w \in L_\lambda$, a $v \in \Sigma^*$ such that for all $i \in \{1, 2, \ldots, k\}$ it holds $|v|_{a_i} \leq m_i$, and a $z \in \left(\Sigma \setminus \Sigma_{q_{[wv]_{\sim L}}}\right)^{*(\Sigma^*, \Sigma, \lambda)} \setminus \{\lambda\}$ such that $\left(q_{[\lambda]_{\sim L}}, y\right) \circlearrowright^* \left(q_{[wv]_{\sim L}}, z\right)$. By Lemma 6.1 there is a $y' \in \Sigma^*$ such that $y'z$ is a permutation of y and $\left(q_{[\lambda]_{\sim L}}, y'\right) \vdash^* \left(q_{[wv]_{\sim L}}, \lambda\right)$. We get $y' \sim_L wv$. Set

$$U = \bigcup_{t \in \Sigma^*} \left\{ u \in \Sigma^* \mid ut \in \mathsf{perm}(v) \wedge wu \in L_\lambda \right\}.$$

We have $\lambda \in U$. Let $u_0 \in U$ such that $|u_0| = \max\left(\left\{ |u| \mid u \in U \right\}\right)$ and let $t_0 \in \Sigma^*$ such that $u_0 t_0 \in \mathsf{perm}(v)$. It follows that for all $i \in \{1, 2, \ldots, k\}$ we have $|t_0|_{a_i} \leq |v|_{a_i} \leq m_i$. Moreover, there exists no $\vec{x} \in M\left([wu_0]_{\sim L}\right)$ with $\vec{x} \leq \psi(t_0)$. Otherwise, we would have an $x' \in \psi^{-1}(\{\vec{x}\})$ which is a non-empty sub-word of t_0 such that $wu_0 x' \in L$, which implies $u_0 x' \in U$. However, this is a contradiction to the maximality of $|u_0|$. That shows that there is no $\vec{x} \in M\left([wu_0]_{\sim L}\right)$ with $\vec{x} \leq \psi(t_0)$. Let now $\vec{x_0} \in M\left([wu_0]_{\sim L}\right)$. There exists a $j \in \{1, 2, \ldots, k\}$ with $|t_0|_{a_j} < \pi_j(\vec{x_0}) \leq m_j$. Since $z \in$

$\left(\Sigma \setminus \Sigma_{q_{[wu_0t_0]\sim_L}}\right)^{*(\Sigma^*,\Sigma,\lambda)} \setminus \{\lambda\}$ and for all $i \in \{1,2,\ldots,k\}$ it holds that $|t_0|_{a_i} \le m_i$, we get $|z|_{a_j} = 0$. That gives $|t_0z|_{a_j} < \pi_j(\vec{x_0})$ and that $\vec{x_0} \le \psi(t_0z)$ is false. So, we have shown $\psi(t_0z) \notin S\left([wu_0]_{\sim_L}\right)$, which implies $wu_0t_0z \notin L$. From $wu_0t_0z \sim_L wvz \sim_L y'z \sim_L y$ it follows that $y \notin L$.

This shows $L_R(A) = L$. $\qquad\qquad\square$

Example 6.5. Consider the language

$$L = \{\, w \in \{a,b\}^* \mid |w|_b = 0 \lor |w|_b = |w|_a \,\}.$$

For $n,m \ge 0$ with $n \ne m$ it holds $a^n \not\sim_L a^m$. Thus, L is not semirecognizable. Theorem 6.4 tells us that L is not accepted by a right one-way jumping finite automaton.

Using the permutation closed semilinear language from Example 4.1 or 6.5, we can answer the open problem from Chigahara *et al.* if **ROWJ** contains all languages from **JFA** [14]:

Corollary 6.6. *The families* **ROWJ** *and* **JFA** *are incomparable.* $\quad\square$

Chigahara *et al.* already showed that $\{\, w \in \{a,b,c\}^* \mid |w|_a = |w|_b = |w|_c \,\}$, which is not context free, is in **ROWJ** [14]. Since the languages from Examples 4.1 and 6.5 are deterministic context free, we get:

Corollary 6.7. *The family* **pROWJ** *is incomparable to* **pCF** *and to* **pDCF**. $\quad\square$

If we do not demand our languages to be permutation closed, acceptance by a right one-way jumping finite automaton does not imply semirecognizability:

Proposition 6.8. *There is an* $L \in$ **ROWJ**$_1$ *which is not semirecognizable.*

Proof. Let A be the DFA

$$(\{q_0,q_1,q_2\},\{a,b\},\delta,\{q_0\},\{q_1\})$$

where $\delta(q_0, a) = q_0$, $\delta(q_0, b) = q_1$, $\delta(q_1, a) = q_2$, $\delta(q_2, b) = q_1$, and δ is un-defined for the non-mentioned arguments. For $n \geq 0$ we have $(q_0, a^n b) \circlearrowright^n$ $(q_0, b) \circlearrowright (q_1, \lambda)$, which implies $a^n b \in L_R(A)$. Furthermore, for $n, m \geq 1$ it holds

$$(q_0, b^m a^n b) \circlearrowright (q_1, b^{m-1} a^n b) \circlearrowright^m (q_2, a^{n-1} b^m) \circlearrowright^n (q_1, b^{m-1} a^{n-1}),$$

which gives us that $b^m a^n b \in L_R(A)$ if and only if $n = m$. So, for $n, m \geq 1$ with $n \neq m$ we get $a^n b \not\sim_{L_R(A)} a^m b$. Hence, $L_R(A)$ is not semirecognizable. $\square$

Theorem 6.4 gives us a criterion when the complement of a language from the class $\mathbf{pROWJ}$ is also in $\mathbf{pROWJ}$:

Corollary 6.9. *For every alphabet Σ and every language $L \in \mathbf{pROWJ}$ over Σ the condition $\Sigma^* \setminus L \in \mathbf{pROWJ}$ holds if and only if L is regular.*

Proof. Let Σ be an alphabet and $L \in \mathbf{pROWJ}$ be a language over Σ. By Theorem 6.4 the language L is semirecognizable, so $|L/\sim_L| < \infty$. Per definition the relations $\sim_L$ and $\sim_{\Sigma^* \setminus L}$ are the same. Theorem 6.4 tells us that $\Sigma^* \setminus L \in \mathbf{pROWJ}$ if and only if

$$|(\Sigma^* \setminus L)/\sim_L| < \infty \Leftrightarrow |(\Sigma^* \setminus L)/\sim_L| + |L/\sim_L| < \infty$$
$$\Leftrightarrow |\Sigma^*/\sim_L| < \infty \Leftrightarrow L \in \mathbf{REG}.$$

$\square$

6.2 Closure Properties of the Languages Accepted by ROWJFAs

We investigate closure properties of $\mathbf{ROWJ}$. The languages from Examples 4.1 and 6.5 show:

Corollary 6.10. *The family $\mathbf{ROWJ}$ is* not *closed under union, union with regular languages, complementation, and permutation closure.* $\square$

Chigahara *et al.* showed that **ROWJ** is not closed under concatenation, not even under concatenation with regular languages from the right [14]. We also get:

Proposition 6.11. *The family* **ROWJ** *is* not *closed under concatenation with regular languages from the left.*

Proof. Consider the regular language $L_1 = \{ w \in \{a, b\}^* \mid |w|_b = 0 \}$ and the language $L_2 = \{ w \in \{a, b\}^* \mid |w|_a = |w|_b \}$, which is in **ROWJ**. Assume that there is a DFA $A = (Q, \{a, b\}, \delta, \{s\}, F)$ with $L_R(A) = L_1 L_2$. For each $n \geq 0$ there is exactly one $q_n \in F$ with $(s, a^n) \vdash^* (q_n, \lambda)$. There are $n, m \in \mathbb{N}$ with $n < m$ and $q_n = q_m$. Since $a^m b^m \in L_1 L_2$, there exists $f \in F$ with $(s, a^m b^m) \vdash^* (q_m, b^m) \vdash^* (f, \lambda)$. This implies that $(s, a^n b^m) \vdash^* (q_m, b^m) \vdash^* (f, \lambda)$, which gives us $a^n b^m \in L_1 L_2$. That is a contradiction because of $m > n$. Thus, $L_1 L_2$ is not in **ROWJ**. □

If we add the condition that the regular language has to be prefix-free, we get a positive closure result:

Proposition 6.12. *The family* **ROWJ** *is closed under concatenation with prefix-free regular languages from the left.*

Proof. For an alphabet Σ, let $L_1 \subseteq \Sigma^*$ be prefix-free and regular and $L_2 \subseteq \Sigma^*$ be in **ROWJ**. If $\lambda \in L_1$, we have $L_1 = \{\lambda\}$ and therefore $L_1 L_2 = L_2$. Thus, assume form now on that $\lambda \notin L_1$. Let $A_1 = (Q_1, \Sigma, \delta_1, \{s_1\}, F_1)$ be a DFAwttf such that $L(A_1) = L_1$. Moreover, let $A_2 = (Q_2, \Sigma, \delta_2, \{s_2\}, F_2)$ be a DFA with $L_R(A_2) = L_2$ and assume that $Q_1 \cap Q_2 = \emptyset$ without loss of generality. Consider the DFA

$$B = ((Q_1 \setminus F_1) \cup Q_2, \Sigma, \delta, \{s_1\}, F_2),$$

where δ is defined as follows. For $(q, a) \in (Q_1 \setminus F_1) \times \Sigma$, let $\delta(q, a) = \delta_1(q, a)$ if we have $\delta_1(q, a) \notin F_1$, and $\delta(q, a) = s_2$ otherwise. For $(q, a) \in Q_2 \times \Sigma$, the value $\delta(q, a)$ is defined if and only if $\delta_2(q, a)$ is defined. In this case we have $\delta(q, a) = \delta_2(q, a)$. We show that $L_R(B) = L_1 L_2$.

First, let $v \in L_1$ and $w \in L_2$. So, there are $a \in \Sigma$, $p \in Q_1$, $q \in F_1$, and $r \in F_2$ such that $(s_1, v) \vdash^*_{A_1} (p, a) \vdash_{A_1} (q, \lambda)$ and $(s_2, w) \circlearrowright^*_{A_2} (r, \lambda)$.

Because L_1 is prefix-free, there are no $x \in \Sigma^* \setminus \{\lambda\}$ and $q' \in F_1$ such that $(s_1, v) \vdash_{A_1}^* (q', x)$. This gives us

$$(s_1, vw) \vdash_B^* (p, aw) \vdash_B (s_2, w) \circlearrowleft_B^* (r, \lambda),$$

which implies $vw \in L_R(B)$.

Now let $v \in L_R(B)$. Since δ_1 is a total function, there are $a \in \Sigma$, $w, x \in \Sigma^*$, $p \in Q_1 \setminus F_1$, and $q \in F_2$ such that $v = wax$ and

$$(s_1, wax) \vdash_B^* (p, ax) \vdash_B (s_2, x) \circlearrowleft_B^* (q, \lambda).$$

So, there is an $r \in F_1$ with $(s_1, wa) \vdash_{A_1}^* (p, a) \vdash_{A_1} (r, \lambda)$ and $(s_2, x) \circlearrowleft_{A_2}^* (q, \lambda)$. This gives us $wa \in L_1$ and $x \in L_2$. $\qquad\square$

We now turn to homomorphisms:

Proposition 6.13. *The family* **ROWJ** *is* not *closed under λ-free homomorphism.*

Proof. Consider the permutation closed language

$$L = \left\{ w \in \{a, b, c\}^* \mid |w|_b = |w|_c \wedge |w|_a \cdot |w|_b = 0 \right\}.$$

It holds $L/\sim_L = \{[\lambda], [a], [bc]\}$, which implies $L \in$ **ROWJ**. Let the λ-free homomorphism $h : \{a, b, c\}^* \to \{a, b\}^*$ be defined by $h(a) = a$, $h(b) = b$, and $h(c) = a$. Then, we get

$$h(L) = \left\{ w \in \{a, b\}^* \mid |w|_b = 0 \vee |w|_b = |w|_a \right\},$$

which was shown to be not in **ROWJ** in Example 6.5. $\qquad\square$

For the operation of inverse homomorphism we have:

Proposition 6.14. *The family* **ROWJ** *is* not *closed under inverse homomorphism.*

Proof. Let $\Sigma = \{a, b, c\}$ and A be the DFA $(\{q_0, q_1, q_2\}, \Sigma, \delta, \{q_0\}, \{q_0, q_2\})$, where δ consists of the rules $q_0 c \to q_0$, $q_0 b \to q_1$, $q_1 a \to q_2$, and $q_2 b \to q_1$.

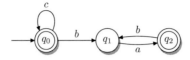

Figure 6.1: The DFA A satisfying $L_R(A) \cap \{ac, b\}^{*(\Sigma^*, \cdot_\Sigma, \lambda_\Sigma)} = \{(ac)^n b^n \mid n \geq 0\}$.

The DFA A is depicted in Figure 6.1. Let $h : \{a, b\}^* \to \{a, b, c\}^*$ be the homomorphism given by $h(a) = ac$ and $h(b) = b$. Furthermore, let $w \in L_R(A) \cap \{ac, b\}^{*(\Sigma^*, \cdot_\Sigma, \lambda_\Sigma)} \setminus \{\lambda\}$. When A interpreted as a right one-way jumping finite automaton reads w, it reaches the first occurrence of the symbol b in state q_0. After reading this b, the automaton is in state q_1. Now, no more c can be read. So, we get $w \in \{ac\}^{*(\Sigma^*, \cdot_\Sigma, \lambda_\Sigma)} \{b\}^{*(\Sigma^*, \cdot_\Sigma, \lambda_\Sigma)}$. Whenever A is in state q_2, it has read the same number of a's and b's. This gives us $w \in \{(ac)^n b^n \mid n > 0\}$. For $n > 0$ we have

$$(q_0, (ac)^n b^n) \circlearrowright^{2n} (q_0, b^n a^n) \circlearrowright^{n+1} (q_2, a^{n-1} b^{n-1}) \circlearrowright^* (q_2, \lambda).$$

This implies

$$L_R(A) \cap \{ac, b\}^{*(\Sigma^*, \cdot_\Sigma, \lambda_\Sigma)} = \{(ac)^n b^n \mid n \geq 0\}.$$

We get

$$h^{-1}(L_R(A)) = h^{-1}\left(L_R(A) \cap h(\{a, b\}^*)\right)$$
$$= h^{-1}\left(L_R(A) \cap \{ac, b\}^{*(\Sigma^*, \cdot_\Sigma, \lambda_\Sigma)}\right)$$
$$= h^{-1}(\{(ac)^n b^n \mid n \geq 0\}) = \{a^n b^n \mid n \geq 0\},$$

which is not in **ROWJ**. $\square$

Next, we consider closure properties of **pROWJ**. To prove the non-closure results from Proposition 2.36, Meduna and Zemek gave examples where the operand languages are in **pROWJ** but the resulting language is not even permutation closed [54]. Hence, it holds:

Proposition 6.15. *The family* **pROWJ** *is* not *closed under intersection with regular languages, Kleene star, Kleene plus, concatenation, λ-free homomorphism, and substitution.* $\square$

The family of permutation closed languages is closed under union, intersection, complementation, and inverse homomorphism. However, Examples 4.1 and 6.5 give us:

Corollary 6.16. *The family* **pROWJ** *is* not *closed under union and complementation.* ☐

Proposition 4.12 and Theorem 6.4 give us:

Corollary 6.17. *Let* Σ *and* Γ *be alphabets,* $L_1, L_2 \in 2^{\Sigma^*}$, $h : \Gamma^* \to \Sigma^*$ *be a monoid homomorphism, and* $n, m \in \mathbb{N}$ *such that* $L_1 \in$ **pROWJ**$_n$ *and* $L_2 \in$ **pROWJ**$_m$. *Then, it holds* $L_1 \cap L_2 \in$ **pROWJ**$_{nm}$ *and also* $h^{-1}(L_1) \in$ **pROWJ**$_n$. ☐

Closure properties of the language families **ROWJ** and **pROWJ** are given in Table 6.1, the corresponding closure properties of **JFA** are also given for comparison.

6.3 Concatenations of Languages

We investigate when a language that is given as the concatenation of a language and a word is in **ROWJ**.

Proposition 6.18. *Let* Σ *be an alphabet,* $w \in \Sigma^*$, *and* $L \subseteq \Sigma^*$. *Then,* $\{w\}L \in$ **ROWJ** *if and only if* $L \in$ **ROWJ**.

Proof. If $L \in$ **ROWJ**, then $\{w\}L \in$ **ROWJ** because of Proposition 6.12. Now assume that $\{w\}L \in$ **ROWJ** and $L \neq \emptyset$. We may also assume that $|w| = 1$. The general case follows from this special case *via* a trivial induction over the length of w. Thus, let $a \in \Sigma$ and let $A = (Q, \Sigma, \delta, \{s\}, F)$ be a DFA with $L_R(A) = \{a\}L$. In the following, we show *via* a contradiction that the value $\delta(s, a)$ is defined. Assume that $\delta(s, a)$ is undefined and let $v \in L$. Because $av \in L_R(A)$, there are a symbol $b \in \Sigma_s$, words $x \in (\Sigma \setminus \Sigma_s)^{*(\Sigma^*, \Sigma, \lambda_\Sigma)}$ and $y \in \Sigma^*$, and a state $p \in F$ such that $v = xby$ and $(s, axby) \circlearrowright^* (\delta(s, b), yax) \circlearrowright^* (p, \lambda)$. This gives us $(s, byax) \vdash$

$(\delta(s,b), yax) \circlearrowright^* (p,\lambda)$, which implies $byax \in L_R(A) = \{a\}L$. This is a contradiction because $b \neq a$. So, $\delta(s,a)$ is defined. Consider the DFA B with $B = (Q, \Sigma, \delta, \{\delta(s,a)\}, F)$. For all $z \in \Sigma^*$ we have $z \in L_R(B)$ if and only if it holds $az \in L_R(A) = \{a\}L$ because of $(s, az) \vdash (\delta(s,a), z)$. That gives us $L_R(B) = L$. □

Next, we give a characterization for concatenations of the form $L\{w\}$, where L is a language and w is a word. To do so, we need the following lemma. It treats the case of a right one-way jumping finite automaton that is only allowed to jump over one of the input symbols.

Lemma 6.19. *Let* $A = (Q, \Sigma, \delta, \{s\}, F)$ *be a DFA and* $a \in \Sigma$ *such that for all* $(q,b) \in Q \times (\Sigma \setminus \{a\})$ *the value* $\delta(q,b)$ *is defined. Then,* $L_R(A)$ *is regular.*

Proof. Consider the DFAwttf

$$B = \left(Q \times (Q \cup \{d\})^Q, \Sigma, \varepsilon, \{(s, \mathrm{id}_{Q \to Q \cup \{d\}})\}, G \right),$$

where $d \notin Q \cup \Sigma$, for all $q \in Q$ we have $\mathrm{id}_{Q \to Q \cup \{d\}}(q) = q$, and

$$G = \left\{ (q, f) \in Q \times (Q \cup \{d\})^Q \mid f(q) \in F \right\}.$$

The function ε is defined as follows:

$$\varepsilon((q,f), b) = \begin{cases} (\delta(q,b), f) & \text{if } b \in \Sigma_{\delta, q}, \\ (q, g \circ f) & \text{otherwise.} \end{cases}$$

The function $g : (Q \cup \{d\}) \to (Q \cup \{d\})$ is defined in the following way:

$$g(p) = \begin{cases} \delta(p, a) & \text{if } p \in Q \text{ and } a \in \Sigma_{\delta, p}, \\ d & \text{otherwise.} \end{cases}$$

This completes the description of B. We show that $L(B) = L_R(A)$.

Let $w \in \Sigma^*$. We decompose the word w into factors that are consumed by A and factors that are jumped over by the automaton: there exist a number $m > 0$, words $w_1, w_2, \ldots, w_m \in \Sigma^*$, symbols $b_2, b_3, \ldots, b_m \in \Sigma \setminus \{a\}$, numbers $n_1, \ldots, n_m \in \mathbb{N}$ with $n_1 n_2 \cdots n_{m-1} > 0$, and $p_1, p_2, \ldots, p_m \in Q$ and $q_2, q_3, \ldots, q_m \in Q$ with

$$
\begin{aligned}
(s, w) &= \left(s, w_1 a^{n_1} \prod_{i=2}^{m} b_i w_i a^{n_i} \right) \\
\vdash_A^{|w_1|} \left(p_1, a^{n_1} \prod_{i=2}^{m} b_i w_i a^{n_i} \right) &\circlearrowright_A^{n_1+1} \left(q_2, w_2 a^{n_2} \left(\prod_{i=3}^{m} b_i w_i a^{n_i} \right) a^{n_1} \right) \\
\vdash_A^{|w_2|} \left(p_2, a^{n_2} \left(\prod_{i=3}^{m} b_i w_i a^{n_i} \right) a^{n_1} \right) &\circlearrowright_A^{n_2+1} \left(q_3, w_3 a^{n_3} \left(\prod_{i=4}^{m} b_i w_i a^{n_i} \right) a^{n_1+n_2} \right)
\end{aligned}
$$

$\ldots$

$$
\begin{aligned}
\vdash_A^{|w_{m-1}|} \left(p_{m-1}, a^{n_{m-1}} b_m w_m a^{n_m + \sum_{i=1}^{m-2} n_i} \right) &\circlearrowright_A^{n_{m-1}+1} \left(q_m, w_m a^{\sum_{i=1}^{m} n_i} \right) \\
\vdash_A^{|w_m|} \left(p_m, a^{\sum_{i=1}^{m} n_i} \right). &
\end{aligned}
$$

We have $w \in L_R(A)$ if and only if $g^{\sum_{i=1}^{m} n_i}(p_m) \in F$. On the other hand, we get

$$
\begin{aligned}
\left((s, \mathrm{id}_{Q \to Q \cup \{d\}}), w \right) &= \left((s, \mathrm{id}_{Q \to Q \cup \{d\}}), w_1 a^{n_1} \prod_{i=2}^{m} b_i w_i a^{n_i} \right) \\
\vdash_B^{|w_1|} \left((p_1, \mathrm{id}_{Q \to Q \cup \{d\}}), a^{n_1} \prod_{i=2}^{m} b_i w_i a^{n_i} \right) &\vdash_B^{n_1+1} \left((q_2, g^{n_1}|_Q), w_2 a^{n_2} \prod_{i=3}^{m} b_i w_i a^{n_i} \right) \\
\vdash_B^{|w_2|} \left((p_2, g^{n_1}|_Q), a^{n_2} \prod_{i=3}^{m} b_i w_i a^{n_i} \right) &\vdash_B^{n_2+1} \left((q_3, g^{n_1+n_2}|_Q), w_3 a^{n_3} \prod_{i=4}^{m} b_i w_i a^{n_i} \right)
\end{aligned}
$$

$\ldots$

$$
\begin{aligned}
\vdash_B^{|w_{m-1}|} \left(\left(p_{m-1}, g^{\sum_{i=1}^{m-2} n_i}|_Q \right), v \right) &\vdash_B^{n_{m-1}+1} \left(\left(q_m, g^{\sum_{i=1}^{m-1} n_i}|_Q \right), w_m a^{n_m} \right) \\
\vdash_B^{|w_m|} \left(\left(p_m, g^{\sum_{i=1}^{m-1} n_i}|_Q \right), a^{n_m} \right), &
\end{aligned}
$$

where $v = a^{n_{m-1}} b_m w_m a^{n_m}$. Set

$$
k = \max \left(\{ r \in \{0, 1, \ldots, n_m\} \mid g^r(p_m) \in Q \} \right).
$$

That gives

$$
\begin{aligned}
\left(\left(p_m, g^{\sum_{i=1}^{m-1} n_i}|_Q \right), a^{n_m} \right) &\vdash_B^{k} \left(\left(g^k(p_m), g^{\sum_{i=1}^{m-1} n_i}|_Q \right), a^{n_m-k} \right) \\
&\vdash_B^{n_m-k} \left(\left(g^k(p_m), g^{\sum_{i=1}^{m} n_i-k}|_Q \right), \lambda \right).
\end{aligned}
$$

Thus, we have $w \in L(B)$ if and only if

$$g^{\sum_{i=1}^{m} n_i}(p_m) = g^{\sum_{i=1}^{m} n_i - k}\left(g^k(p_m)\right) \in F,$$

which holds if and only if $w \in L_R(A)$. That shows $L(B) = L_R(A)$ and that $L_R(A)$ is a regular language. $\qquad\square$

Our characterization for languages of the form $L\{w\}$ generalizes a result from Chigahara *et al.* which says that the language $\{\, va \mid v \in \{a, b\}^* \wedge |v|_a = |v|_b \,\}$ is not in **ROWJ** [14]:

Theorem 6.20. *Let Σ be an alphabet, $w \in \Sigma^* \setminus \{\lambda\}$, and $L \subseteq \Sigma^*$. Then, $L\{w\} \in$ **ROWJ** if and only if L is regular.*

Proof. If L is regular, then $L\{w\}$ is also regular, which means that $L\{w\}$ is in **ROWJ**. Assume now, that $L\{w\}$ is in **ROWJ**. As in the proof of Proposition 6.18, we can assume that there is an $a \in \Sigma$ with $w = a$. Let $A = (Q, \Sigma, \delta, \{s\}, F)$ be a DFA with $L_R(A) = L\{a\}$. Consider the DFA $B = (Q \cup \{d\}, \Sigma, \varepsilon, \{s\}, F)$ with $d \notin Q \cup \Sigma$. The map ε is defined as follows: For $(q, b) \in Q \times \Sigma$ we set $\varepsilon(q, b) = \delta(q, b)$ if $\delta(q, b)$ is defined. For $(q, b) \in Q \times (\Sigma \setminus \{a\})$ we define $\varepsilon(q, b) = d$ if $\delta(q, b)$ is undefined. For all $q \in Q$ the value $\varepsilon(q, a)$ is undefined if $\delta(q, a)$ is undefined. Finally, for all $b \in \Sigma$ it holds $\varepsilon(d, b) = d$. By Lemma 6.19, $L_R(B)$ is regular. We show that $L_R(A) = L_R(B)$. Then, the regularity of $L_R(A) = L\{a\}$ implies the regularity of L.

First, let $v \in L_R(B)$ and $f \in F$ with $(s, v) \circlearrowright_B^* (f, \lambda)$. For a state $q \in Q$, a $b \in \Sigma_{\varepsilon, q}$, and words $x \in (\Sigma \setminus \Sigma_{\varepsilon, q})^{*(\Sigma^*, \Sigma, \lambda_\Sigma)}$ and $y \in \Sigma^*$ with $(s, v) \circlearrowright_B^*$ (q, xby) we have $x \in (\Sigma \setminus \Sigma_{\delta, q})^{*(\Sigma^*, \Sigma, \lambda_\Sigma)}$ and $(q, xby) \circlearrowright_B^{|xb|} (\varepsilon(q, b), yx) \circlearrowright_B^*$ (f, λ). This implies $\varepsilon(q, b) \neq d$, which tells us $b \in \Sigma_{\delta, q}$ and $\delta(q, b) = \varepsilon(q, b)$. We get $(q, xby) \circlearrowright_A^{|xb|} (\varepsilon(q, b), yx)$. By induction, we see that $(s, v) \circlearrowright_A^*$ (f, λ). Therefore, we have $v \in L_R(A)$.

Now, let $v \in L_R(A)$ and $f \in F$ with $(s, v) \circlearrowright_A^* (f, \lambda)$. Assume that $v \notin L_R(B)$. Then, there exists a symbol out of $\Sigma \setminus \{a\}$ that is jumped over during the processing of A, when the starting configuration is (s, v). The part of v that is visited by A before it jumps over the first symbol out of $\Sigma \setminus \{a\}$ is decomposed into factors that are consumed by A and factors that are jumped over by the device under consideration: there are a natural number $m > 0$, words $w_1, w_2, \ldots, w_{m+2} \in \Sigma^*$, symbols $b_2, b_3, \ldots, b_{m+1} \in \Sigma \setminus \{a\}$ and $c \in \Sigma$, numbers $n_1, \ldots, n_m \in \mathbb{N}$ with $n_1 n_2 \cdots n_{m-1} > 0$, and $p_1, p_2, \ldots, p_m, q_2, q_3, \ldots, q_{m+1} \in Q$ such that for every number $i \in \{1, 2, \ldots, m-1\}$ it holds $b_{i+1} \in \Sigma_{p_i}$ and such that $a^{n_m} b_{m+1} w_{m+1} \in (\Sigma \setminus \Sigma_{\delta, p_m})^{*(\Sigma^*, \cdot_\Sigma, \lambda_\Sigma)}$, $c \in \Sigma_{\delta, p_m}$, and

$$(s, v) = \left(s, w_1 a^{n_1} \left(\prod_{i=2}^{m} b_i w_i a^{n_i}\right) z\right)$$

$$\vdash_A^{|w_1|} \left(p_1, a^{n_1} \left(\prod_{i=2}^{m} b_i w_i a^{n_i}\right) z\right) \circlearrowleft_A^{n_1+1} \left(q_2, w_2 a^{n_2} \left(\prod_{i=3}^{m} b_i w_i a^{n_i}\right) z a^{n_1}\right)$$

$$\vdash_A^{|w_2|} \left(p_2, a^{n_2} \left(\prod_{i=3}^{m} b_i w_i a^{n_i}\right) z a^{n_1}\right) \circlearrowleft_A^{n_2+1} \left(q_3, w_3 a^{n_3} \left(\prod_{i=4}^{m} b_i w_i a^{n_i}\right) z a^{n_1+n_2}\right)$$

$\ldots$

$$\vdash_A^{|w_{m-1}|} \left(p_{m-1}, a^{n_{m-1}} b_m w_m a^{n_m} z a^{\sum_{i=1}^{m-2} n_i}\right) \circlearrowleft_A^{n_{m-1}+1} \left(q_m, w_m a^{n_m} z a^{\sum_{i=1}^{m-1} n_i}\right)$$

$$\vdash_A^{|w_m|} \left(p_m, a^{n_m} z a^{\sum_{i=1}^{m-1} n_i}\right) \circlearrowleft_A^* \left(q_{m+1}, w_{m+2} a^{\sum_{i=1}^{m} n_i} b_{m+1} w_{m+1}\right)$$

$$\circlearrowleft_A^* (f, \lambda),$$

where $z = b_{m+1} w_{m+1} c w_{m+2}$. We get

$$\left(s, w_1 \left(\prod_{i=2}^{m} b_i w_i\right) w_{m+1} c w_{m+2} a^{\sum_{i=1}^{m} n_i} b_{m+1}\right)$$
$$\vdash_A^{|w_1 \prod_{i=2}^{m} b_i w_i|} \left(p_m, w_{m+1} c w_{m+2} a^{\sum_{i=1}^{m} n_i} b_{m+1}\right)$$
$$\circlearrowleft_A^{|w_{m+1} c|} \left(q_{m+1}, w_{m+2} a^{\sum_{i=1}^{m} n_i} b_{m+1} w_{m+1}\right) \circlearrowleft_A^* (f, \lambda).$$

This implies

$$w_1 \left(\prod_{i=2}^{m} b_i w_i\right) w_{m+1} c w_{m+2} a^{\sum_{i=1}^{m} n_i} b_{m+1} \in L_R(A) = L\{a\},$$

a contradiction. Hence, it holds $v \in L_R(B)$. □

We consider the case of two languages over disjoint alphabets:

Proposition 6.21. *Let Σ, Σ_1, and Σ_2 be alphabets with $\Sigma_1 \cap \Sigma_2 = \emptyset$ and $\Sigma = \Sigma_1 \cup \Sigma_2$. Furthermore, let $L_1 \subseteq \Sigma_1^{*(\Sigma^*, \cdot \Sigma, \lambda_\Sigma)}$ and $L_2 \subseteq \Sigma_2^{*(\Sigma^*, \cdot \Sigma, \lambda_\Sigma)}$ with $L_1 \neq \emptyset \neq L_2 \neq \{\lambda\}$ and $L_1 L_2 \in \mathbf{ROWJ}$. Then, L_1 is regular and $L_2 \in \mathbf{ROWJ}$.*

Proof. The proof is similar to the proof of Proposition 6.18. Let $A = (Q, \Sigma, \delta, \{s\}, F)$ be a DFA such that it holds that $L_R(A) = L_1 L_2$. Let $m \geq 0$, $a_1, a_2, \cdots, a_m \in \Sigma_1$, and also $w = a_1 a_2 \cdots a_m \in L_1$. We show by induction that for each $n \in \{0, 1, \ldots, m\}$ there is a state $q_n \in Q$ with $(s, w) \vdash^n (q_n, a_{n+1} a_{n+2} \cdots a_m)$. For $n = 0$, we just set $q_0 = s$. Assume that for a fixed $k \in \{0, 1, \ldots, m-1\}$ we already know that there is a state $q_k \in Q$ with $(s, w) \vdash^k (q_k, a_{k+1} a_{k+2} \cdots a_m)$. If the value $\delta(q_k, a_{k+1})$ is defined, we have

$$(s, w) \vdash^{k+1} (\delta(q_k, a_{k+1}), a_{k+2} a_{k+3} \cdots a_m).$$

Therefore, let now $\delta(q_k, a_{k+1})$ be undefined and let $v \in L_2 \setminus \{\lambda\}$. We get $(s, wv) \vdash^k (q_k, a_{k+1} a_{k+2} \cdots a_m v)$. Because of $wv \in L_R(A)$, there exist a symbol $b \in \Sigma_{q_k}$, words $x \in (\Sigma \setminus \Sigma_{q_k})^{*(\Sigma^*, \cdot \Sigma, \lambda_\Sigma)}$ and $y \in \Sigma^*$, and a state $p \in F$ such that $a_{k+2} a_{k+3} \cdots a_m v = xby$ and

$$(q_k, a_{k+1} xby) \circlearrowright^{|a_{k+1} xb|} (\delta(q_k, b), y a_{k+1} x) \circlearrowright^* (p, \lambda).$$

This implies

$$(s, a_1 a_2 \cdots a_k xby a_{k+1}) \vdash^k (q_k, xby a_{k+1}) \circlearrowright^{|xb|} (\delta(q_k, b), y a_{k+1} x) \circlearrowright^* (p, \lambda),$$

which gives

$$a_1 a_2 \cdots a_k a_{k+2} a_{k+3} \cdots a_m v a_{k+1} = a_1 a_2 \cdots a_k xby a_{k+1} \in L_R(A).$$

We have

$$a_1 a_2 \cdots a_k a_{k+2} a_{k+3} \cdots a_m v a_{k+1} \in \Sigma^* \setminus \left(\Sigma_1^{*(\Sigma^*, \cdot \Sigma, \lambda_\Sigma)} \Sigma_2^{*(\Sigma^*, \cdot \Sigma, \lambda_\Sigma)} \right) \subseteq \Sigma^* \setminus (L_1 L_2),$$

which is a contradiction. So, the value $\delta(q_k, a_{k+1})$ has to be defined and we have shown by induction that for each $n \in \{0, 1, \ldots, m\}$ there is a state $q_n \in Q$ with $(s, w) \vdash^n (q_n, a_{n+1}a_{n+2} \cdots a_m)$. We set $q_w = q_m$ and get $(s, w) \vdash^{|w|} (q_w, \lambda)$.

Let $w \in L_1$. We consider the DFA $B_w = (Q, \Sigma, \delta_2, \{q_w\}, F)$, where for all $(q, a) \in Q \times \Sigma_1$ the value $\delta_2(q, a)$ is undefined and for all $(q, a) \in Q \times \Sigma_2$ we have $\delta_2(q, a) = \delta(q, a)$. For every $v \in \Sigma^*$ it holds

$$v \in L_R(B_w) \Leftrightarrow \left(\exists f \in F : (q_w, v) \circlearrowright^*_{B_w} (f, \lambda) \right)$$
$$\Leftrightarrow \left(v \in \Sigma_2^{*(\Sigma^*, \cdot_\Sigma, \lambda_\Sigma)} \wedge \exists f \in F : (s, wv) \circlearrowright^*_A (f, \lambda) \right)$$
$$\Leftrightarrow \left(v \in \Sigma_2^{*(\Sigma^*, \cdot_\Sigma, \lambda_\Sigma)} \wedge wv \in L_R(A) = L_1 L_2 \right) \Leftrightarrow v \in L_2.$$

This shows $L_R(B_w) = L_2$, so L_2 is in **ROWJ**.

Let $v \in L_2$. We set $Q_v = \{ q \in Q \mid \exists f \in F : (q, v) \circlearrowright^*_A (f, \lambda) \}$ and define the DFA C_v to be $(Q, \Sigma, \delta_1, \{s\}, Q_v)$, where for all $(q, a) \in Q \times \Sigma_2$ the value $\delta_1(q, a)$ is undefined and for all $(q, a) \in Q \times \Sigma_1$ we have $\delta_1(q, a) = \delta(q, a)$. For every $w \in \Sigma^*$ we get

$$w \in L(C_v) \Leftrightarrow \left(\exists q \in Q_v : (s, w) \vdash^*_{C_v} (q, \lambda) \right)$$
$$\Leftrightarrow \left(w \in \Sigma_1^{*(\Sigma^*, \cdot_\Sigma, \lambda_\Sigma)} \wedge \exists q \in Q, f \in F : (s, wv) \vdash^*_A (q, v) \circlearrowright^*_A (f, \lambda) \right),$$

which is equivalent to $w \in L_1$. That gives $L(C_v) = L_1$, so L_1 is regular. □

Adding prefix-freeness for L_1, we get an equivalence by Proposition 6.12:

Proposition 6.22. *Let Σ, Σ_1, and Σ_2 be alphabets fulfilling that $\Sigma_1 \cap \Sigma_2 = \emptyset$ and $\Sigma = \Sigma_1 \cup \Sigma_2$. Furthermore, let $L_1 \subseteq \Sigma_1^{*(\Sigma^*, \cdot_\Sigma, \lambda_\Sigma)}$ be prefix-free and $L_2 \subseteq \Sigma_2^{*(\Sigma^*, \cdot_\Sigma, \lambda_\Sigma)}$ such that it holds $L_1 \neq \emptyset \neq L_2 \neq \{\lambda\}$. Then, we have $L_1 L_2 \in$ **ROWJ** if and only if both $L_1 \in$ **REG** and $L_2 \in$ **ROWJ** hold.* □

Closed under	Language family		
	ROWJ	**JFA**	**pROWJ**
Union	no	yes	no
Union with reg. lang.	no	no	no
Intersection	no	yes	yes
Intersection with reg. lang.	no	no	no
Complementation	no	yes	no
Reversal	no	yes	yes
Kleene star	no	no	no
Kleene plus	no	no	no
Concatenation	no	no	no
Right conc. with reg. lang.	no	no	no
Left conc. with reg. lang.	no	no	no
Left conc. with prefix-free reg. lang.	yes	no	no
Homomorphism	no	no	no
λ-free homomorphism	no	no	no
Inverse homomorphism	no	yes	yes
Substitution	no	no	no
Permutation closure	no	yes	yes

Table 6.1: Closure properties of the language families **ROWJ**, **JFA**, and **pROWJ**.

7 Problems Involving Right One-Way Jumping Finite Automata

We investigate the decidability and complexity of problems involving right one-way jumping finite automata. It is easy to see that for a DFA A with input alphabet Σ we have $L_R(A) = \emptyset \Leftrightarrow L(A) = \emptyset$, $|L_R(A)| < \infty \Leftrightarrow |L(A)| < \infty$, and $L_R(A) = \Sigma^* \Leftrightarrow L(A) = \Sigma^*$. So, for right one-way jumping finite automata emptiness, finiteness, and universality are **NL**-complete. It is open if regularity, context-freeness, disjointness, inclusion, and equivalence are decidable for right one-way jumping finite automata. Also, it is not known whether there is an algorithm that minimizes the number of states of a right one-way jumping finite automaton. For a DFA A the problem to decide whether $L_R(A) = L(A)$ and the problem whether $L_R(A)$ is letter bounded, are both **NL**-complete. For a fixed $n > 0$ and a one-turn PDA A it is undecidable whether $L(A)$ is semirecognizable, whether $L(A) \in$ **ROWJ**, whether $L(A) \in$ **pROWJ**, whether $L(A)$ is n-semirecognizable, whether $L(A) \in$ **ROWJ$_n$**, and whether $L(A) \in$ **pROWJ$_n$**. We consider variants of the word problem for right one-way jumping finite automata by asking whether a given word can be completed by a prefix, suffix, etc., to a word that is accepted by a given right one-way jumping finite automaton. Since the family **ROWJ** is not closed under intersection with regular languages, we cannot deduce decidability of these problems from decidability of emptiness by applying an intersection construction (with a regular set). Nevertheless, by a careful inspection of the computations of right one-way jumping finite automata, it turns out that the considered variants of the word problem are all in **PSPACE**. We also consider decision problems that deal with the relation between a given right one-way jumping finite automaton and a given ordinary jumping finite automaton. As a consequence of these results, we get that, if the input alphabet is fixed and of cardinality at least two, it is **NL**-complete to decide whether a given right one-way jumping finite automaton accepts a permutation closed language. For right one-way jumping finite automata that accept a permutation closed language over a fixed alphabet, the word problem, emptiness, finiteness, universality, regularity, context-freeness, acceptance by a counter automaton, deterministic context-freeness, acceptance by a λ-free deterministic pushdown automaton, acceptance by a deterministic counter automaton, acceptance by a λ-free deterministic counter automaton, disjointness, and inclusion are all in **L**.

7.1 Basic Properties

Emptiness, finiteness, and universality are **NL**-complete for right one-way jumping finite automata:

Proposition 7.1. *Let A be a DFA with input alphabet Σ. Then, the problems to decide whether $L_R(A) = \emptyset$, whether $L_R(A) < \infty$, and whether $L_R(A) = \Sigma^*$, are all **NL**-complete. This also holds, if Σ is fixed with $|\Sigma| \geq 2$.*

Proof. It is well-known that these problems are **NL**-complete if we replace $L_R(A)$ by $L(A)$. Clearly, we have $L_R(A) = \emptyset \Leftrightarrow L(A) = \emptyset$ and $|L_R(A)| < \infty \Leftrightarrow |L(A)| < \infty$. If it holds $L(A) = \Sigma^*$, then $L_R(A) = \Sigma^*$. Now, assume that there is a $w \in \Sigma^* \setminus L(A)$. Let $A = (Q, \Sigma, \delta, \{s\}, F)$. There are two possibilities:

1. There is a $q \in Q \setminus F$ such that $(s, w) \vdash^* (q, \lambda)$. Then, we also have $(s, w) \circlearrowright^* (q, \lambda)$, which gives us $w \notin L_R(A)$.

2. There are a state $q \in Q$, words $x, y \in \Sigma^*$, and an $a \in \Sigma$ with $w = xay$ such that $(s, x) \vdash^* (q, \lambda)$ and $\delta(q, a)$ is undefined. Then, we have $(s, xa) \circlearrowright^* (q, a)$ and get $xa \notin L_R(A)$.

So, we have shown $L_R(A) = \Sigma^* \Leftrightarrow L(A) = \Sigma^*$. $\qquad\square$

We consider the problem if a given DFA accepts the same languages in ordinary mode and in right one-way jumping mode:

Proposition 7.2. *For a DFA A the problem to decide whether $L_R(A) = L(A)$, is* **NL**-*complete. This is also true if the input alphabet is fixed and its cardinality is at least two.*

Proof. Let $A = (Q, \Sigma, \delta, \{s\}, F)$ be a DFA and $C(A)$ be the following decidable condition: there exist a symbol $a \in \Sigma$ and states $q, r \in Q$ such that q is reachable from s, the value $\delta(q, a)$ is undefined, the state r is reachable from q, and $\delta(r, a)$ is defined with $\delta(r, a) \in F$. We show that $C(A)$ holds if and only if $L(A) \subset L_R(A)$.

First, assume $C(A)$ is true with $a \in \Sigma$ and $q, r \in Q$ as above. Let $v, w \in \Sigma^*$ with $(s, v) \vdash^* (q, \lambda)$ and $(q, w) \vdash^* (r, \lambda)$. Then, we have $(s, vaw) \vdash^* (q, aw)$, which gives us that vaw is not in $L(A)$. On the other hand, we get $(s, vaw) \circlearrowright^* (q, aw) \circlearrowright^* (r, a)$, which shows $vaw \in L_R(A)$. So, we have $L(A) \subset L_R(A)$.

Now, assume $L(A) \subset L_R(A)$ and let $x \in L_R(A) \setminus L(A)$. We consider the last jump over a symbol that A interpreted as a right one-way jumping finite automaton does while processing the input x: there exist a symbol $a \in \Sigma$, states $q, r \in Q$, and words $y \in \Sigma^*$ and $z \in (\Sigma \setminus \Sigma_q)^{*(\Sigma^* \cdot \Sigma \cdot \lambda_\Sigma)}$ such that the value $\delta(q, a)$ is undefined, the value $\delta(r, a)$ is defined with $\delta(r, a) \in F$, and we have $(s, x) \circlearrowright^* (q, zay)$ and $(q, yz) \vdash^* (r, \lambda)$. By Lemma 6.1, there is a word $v \in \Sigma^*$ with $(s, v) \vdash^* (q, \lambda)$. So, the condition $C(A)$ is fulfilled. After all, we have shown that $C(A)$ is true if and only if $L(A) \subset L_R(A)$.

Next, we show the containment of $C(A)$ in **NL**. A nondeterministic logspace bounded Turing machine guesses a letter $a \in \Sigma$ and two words $v, w \in \Sigma^*$ in a letter-by-letter fashion that constitute $C(A)$, that is, that there are states $q, r \in Q$ such that $(s, v) \vdash^* (q, \lambda)$, the value $\delta(q, a)$ is undefined, $(q, w) \vdash^* (r, \lambda)$, and $\delta(r, a)$ is defined with $\delta(r, a) \in F$. Since these properties can be decided in **NL**, the condition $C(A)$ is in **NL**.

To show the **NL** lower bound, we give a log-space many-one reduction of a variant of the **NL**-complete graph reachability problem to the condition C for DFAs: Given a directed graph $G = (V, E)$ and two vertices $s, t \in V$

such that every vertex of G has exactly two successors, is there a path from s to t in G? Without loss of generality we may assume that there is an $m \geq 2$ such that $V = \{1, 2, \ldots, m\}$, $s = 1$, and $t = m$. Then, let A be the DFA $(V \cup \{0\}, \{a, b\}, \delta, \{0\}, \{m\})$, where

$$\delta = \{\, ib \to j, ia \to k \mid (i, j), (i, k) \in E \wedge j < k \,\} \cup \{0b \to 1\}.$$

The condition $C(A)$ holds if and only if there is a path from 1 to m in G. It is easy to see that A can be constructed from G in deterministic logarithmic space. This shows NL-hardness of the considered problem. □

For letter boundedness of languages accepted by right one-way jumping finite automata we have:

Proposition 7.3. *For a DFA A the problem to decide whether $L_R(A)$ is letter bounded, is* NL-*complete. This is also true if the input alphabet is fixed and its cardinality is at least two.*

Proof. Let $k \geq 1$ and A be a DFA with input alphabet $\Sigma = \{a_1, a_2, \ldots, a_k\}$. Since $L(A)$ is a Parikh-equivalent sublanguage of $L_R(A)$, we have: The language $L_R(A)$ is letter bounded if and only if on the one hand $L(A) = L_R(A)$ and on the other hand $L(A)$ is letter bounded. The problem whether $L(A) = L_R(A)$ holds, is in NL by Proposition 7.2. The problem whether $L(A)$ is letter bounded, is also in NL because a word in

$$L(A) \setminus \{a_1\}^{*(\Sigma^*, \Sigma, \lambda_\Sigma)} \{a_2\}^{*(\Sigma^*, \Sigma, \lambda_\Sigma)} \cdots \{a_k\}^{*(\Sigma^*, \Sigma, \lambda_\Sigma)}$$

can be guessed in a letter-by-letter fashion in logarithmic space. Hence, the problem whether $L_R(A)$ is letter bounded, is in NL, as well.

The NL-hardness of this problem is shown with the same reduction as in the proof of Proposition 7.2. If there is a path from 1 to m in the graph G from this proof, for the constructed DFA A it holds $L(A) \subset L_R(A)$. Otherwise, we have $L_R(A) = L(A) = \emptyset$. □

We use an adaptation of the approach by Bar-Hillel *et al.* that shows that regularity, universality, and permutation closure are undecidable for one-turn pushdown automata [1] to get undecidability results for semirecognizability and acceptance by right one-way jumping finite automata:

Proposition 7.4. *Let $n > 0$ be fixed. Then, for a one-turn PDA A the following problems are undecidable.*

- *Is $L(A)$ semirecognizable?*

- *Is $L(A) \in \mathbf{ROWJ}$?*

- *Is $L(A) \in \mathbf{pROWJ}$?*

- *Is $L(A)$ n-semirecognizable?*

- *Is $L(A) \in \mathbf{ROWJ_n}$?*

- *Is $L(A) \in \mathbf{pROWJ_n}$?*

Proof. An instance of the Post correspondence problem, PCP for short, is defined as a tuple (Σ, Γ, f, g), where Σ and Γ are disjoint alphabets and $f, g : \Sigma^* \to \Gamma^*$ are monoid homomorphisms. We only consider instances where for all $a \in \Sigma$ it holds $f(a)g(a) \neq \lambda$. Such an instance is called *positive* if there exists a $w \in \Sigma^* \backslash \{\lambda\}$ with $f(w) = g(w)$, otherwise the instance is said to be *negative*. It is undecidable if an instance is positive, see for example the book *Introduction to Formal Language Theory* by Harrison [37]. For disjoint alphabets Σ, Γ and a monoid homomorphism $f : \Sigma^* \to \Gamma^*$ let

$$L_f = \left\{ w(f(w))^R \;\middle|\; w \in \Sigma^* \backslash \{\lambda\} \right\} \subseteq (\Sigma \cup \Gamma)^*,$$

which is effectively accepted by a deterministic one-turn PDA. For disjoint alphabets Σ, Γ and monoid homomorphisms $f, g : \Sigma^* \to \Gamma^*$ set

$$L_{f,g} = ((\Sigma \cup \Gamma)^* \backslash L_f) \cup ((\Sigma \cup \Gamma)^* \backslash L_g).$$

Because deterministic one-turn PDA languages are effectively closed under complement and one-turn PDA languages are effectively closed under union, we can construct a one-turn PDA that accepts $L_{f,g}$. By De Morgan, we have $L_{f,g} = (\Sigma \cup \Gamma)^* \backslash (L_f \cap L_g)$. So, an instance (Σ, Γ, f, g) of the PCP is negative if and only if $L_{f,g} = (\Sigma \cup \Gamma)^*$. The language $(\Sigma \cup \Gamma)^*$ is in $\mathbf{pROWJ_1}$.

Let now (Σ, Γ, f, g) be a positive instance of the PCP and $w \in \Sigma^* \backslash \{\lambda\}$ with $f(w) = g(w)$. For all $n, m \in \mathbb{N}$ with $n \neq m$ it holds $w^n \not\sim_{L_{f,g}} w^m$.

Hence, $L_{f,g}$ is not semirecognizable. We show by contradiction that $L_{f,g}$ is not in **ROWJ**. Assume that the device $A = (Q, \Sigma \cup \Gamma, \delta, \{s\}, F)$ is a DFA with $L_R(A) = L_{f,g}$. For $n, m \in \mathbb{N}_{>0}$ it holds $w^n \left((f(w))^R\right)^m \in L_{f,g}$ if and only if $n \neq m$. We consider the following two possible cases:

1. For every $n > 1$ there is a state $q_n \in Q$ with $\left(s, w^n(f(w))^R\right) \circlearrowright^* \left(q_n, (f(w))^R\right)$. This implies that for all $n, m \in \mathbb{N}_{>1}$ we have

$$\left(s, w^n \left((f(w))^R\right)^m\right) \circlearrowright^* \left(q_n, \left((f(w))^R\right)^m\right)$$

   because whenever A is in a state in which it cannot read any symbol that appears in $(f(w))^R$, it also cannot read any symbol that appears in $((f(w))^R)^m$. There are some $n_0, n_1 \in \mathbb{N}_{>1}$ with $n_0 < n_1$ and $q_{n_0} = q_{n_1}$ and an $e \in F$ with

$$\left(s, w^{n_0} \left((f(w))^R\right)^{n_1}\right) \circlearrowright^* \left(q_{n_0}, \left((f(w))^R\right)^{n_1}\right) \circlearrowright^* (e, \lambda).$$

   It follows

$$\left(s, w^{n_1} \left((f(w))^R\right)^{n_1}\right) \circlearrowright^* \left(q_{n_0}, ((f(w))^R))^{n_1}\right) \circlearrowright^* (e, \lambda).$$

   This is a contradiction because $w^{n_1} \left((f(w))^R\right)^{n_1} \notin L_{f,g}$.

2. There is some $n > 1$ for which no $q \in Q$ with $\left(s, w^n(f(w))^R\right) \circlearrowright^* \left(q, (f(w))^R\right)$ exists. It follows that there are $p, r \in Q$, $z \in \Sigma^* \setminus \{\lambda\}$, $x, y \in \Gamma^*$, and $a \in \Gamma$ with $xay = (f(w))^R$ and

$$\left(s, w^n(f(w))^R\right) \circlearrowright^* (p, ayzx) \vdash (r, yzx).$$

   This implies

$$\left(s, w^n \left((f(w))^R\right)^n\right) \circlearrowright^* \left(p, ay \left((f(w))^R\right)^{n-1} zx\right)$$
$$\vdash \left(r, y \left((f(w))^R\right)^{n-1} zx\right).$$

   By Lemma 6.1, there is a $u \in (\Sigma \cup \Gamma)^*$ such that $uy \left((f(w))^R\right)^{n-1} zx$ is a permutation of $w^n \left((f(w))^R\right)^n$ and $(s, u) \vdash^+ (r, \lambda)$. It follows that $|u|_a = 1$. That gives us

$$uy \left((f(w))^R\right)^{n-1} zx \in (\Sigma \cup \Gamma)^* \setminus \left(\Sigma^{*((\Sigma \cup \Gamma)^*, \cdot_{\Sigma \cup \Gamma}, \lambda_{\Sigma \cup \Gamma})} \Gamma^{*((\Sigma \cup \Gamma)^*, \cdot_{\Sigma \cup \Gamma}, \lambda_{\Sigma \cup \Gamma})}\right),$$

which is a subset of $L_{f,g}$. So, there is an $e \in F$ with

$$\left(s, uy\left((f(w))^R\right)^{n-1} zx\right) \vdash^+ \left(r, y\left((f(w))^R\right)^{n-1} zx\right) \circlearrowright^* (e, \lambda).$$

We get

$$\left(s, w^n\left((f(w))^R\right)^n\right) \circlearrowright^+ \left(r, y\left((f(w))^R\right)^{n-1} zx\right) \circlearrowright^* (e, \lambda),$$

which is a contradiction because $w^n\left((f(w))^R\right)^n \notin L_{f,g}$.

That shows that $L_{f,g}$ is not in **ROWJ** if (Σ, Γ, f, g) is a positive instance of the PCP. □

7.2 Variants of the Word Problem

We investigate the problem if we can extend a given word such that it becomes accepted by a given right one-way jumping finite automaton:

Theorem 7.5. *Let A be a DFA with input alphabet Σ and $w \in \Sigma^*$. Then, the following problems are all in* PSPACE.

- *Is there a $v \in L_R(A)$ such that w is a prefix of v?*
- *Is there a $v \in L_R(A)$ such that w is a suffix of v?*
- *Is there a $v \in L_R(A)$ such that w is a factor of v?*
- *Is there a $v \in L_R(A)$ such that w is a sub-word of v?*

Proof. Let $A = (Q, \Sigma, \delta, \{s\}, F)$. We use the following notation. A $c \in ((\Sigma \times \mathbb{N}_{>0}) \cup Q)^*$ is called a *modified configuration* of A if there are a state $q \in Q$, numbers $i, k \in \mathbb{N}$ with $i \leq k$, symbols $a_1, a_2, \ldots, a_k \in \Sigma$, and numbers $n_1, n_2, \ldots, n_k \in \mathbb{N}_{>0}$ with $n_1 < n_2 < \cdots < n_k$ such that

$$c = (a_1, n_1)(a_2, n_2) \cdots (a_i, n_i)q(a_{i+1}, n_{i+1})(a_{i+2}, n_{i+2}) \cdots (a_k, n_k).$$

For $c, q, i, k, a_1, a_2, \ldots, a_k, n_1, n_2, \ldots, n_k$ as in the last sentence we set

- state$(c) = q \in Q$,

- lsym$(c) = a_1 a_2 \cdots a_i \in \Sigma^*$,

- rsym$(c) = a_{i+1} a_{i+2} \cdots a_k \in \Sigma^*$,

- lnum$(c) = \{n_1, n_2, \ldots, n_i\} \subset \mathbb{N}_{>0}$,

- rnum$(c) = \{n_{i+1}, n_{i+2}, \ldots, n_k\} \subset \mathbb{N}_{>0}$, and

- num$(c) = $ lnum$(c) \cup$ rnum(c).

Let mc(A) be the set of all modified configurations of A.

The binary relations $\rightsquigarrow$ and $\downarrow$ on mc(A) are defined as follows: Let $p, q \in Q$, $i, k \in \mathbb{N}$ with $i < k$, $a_1, a_2, \ldots, a_k \in \Sigma$, and $n_1, n_2, \ldots, n_k \in \mathbb{N}_{>0}$ with $n_1 < n_2 < \cdots < n_k$. Then, we have

$$(a_1, n_1)(a_2, n_2) \cdots (a_i, n_i)q(a_{i+1}, n_{i+1})(a_{i+2}, n_{i+2}) \cdots (a_k, n_k)$$
$$\rightsquigarrow (a_1, n_1)(a_2, n_2) \cdots (a_i, n_i)p(a_{i+2}, n_{i+2})(a_{i+3}, n_{i+3})c \ldots (a_k, n_k)$$

if $(qa_{i+1} \rightarrow p) \in \delta$. If $a_{i+1} \notin \Sigma_q$, we set

$$(a_1, n_1)(a_2, n_2) \ldots (a_i, n_i)q(a_{i+1}, n_{i+1})(a_{i+2}, n_{i+2}) \ldots (a_k, n_k)$$
$$\rightsquigarrow (a_1, n_1)(a_2, n_2) \ldots (a_{i+1}, n_{i+1})q(a_{i+2}, n_{i+2})(a_{i+3}, n_{i+3}) \ldots (a_k, n_k).$$

Furthermore, we define

$$(a_1, n_1)(a_2, n_2) \cdots (a_k, n_k)q \downarrow q(a_1, n_1)(a_2, n_2) \cdots (a_k, n_k).$$

For $k \geq 0$, relations $b_1, b_2, \ldots, b_k \in \{\rightsquigarrow, \downarrow\}$, $w = b_1 b_2 \cdots b_k \in \{\rightsquigarrow, \downarrow\}^*$, and $c, d \in$ mc(A) we write $c \searrow^w d$ if there exist modified configurations $c = c_0, c_1, c_2, \ldots, c_k = d$ such that for all $i \in \{0, 1, \ldots, k-1\}$ it holds $c_i \, b_{i+1} \, c_{i+1}$. For $L \subseteq \{\rightsquigarrow, \downarrow\}^*$ and $c, d \in$ mc(A) we write $c \searrow^L d$ if there exists an $x \in L$ with $c \searrow^x d$. With these definitions we get

$$L_R(A) = \Big\{ v \in \Sigma^* \; \Big| \; \exists f \in F : \exists c \in \mathrm{mc}(A) : \Big(\mathrm{lsym}(c) = \lambda$$
$$\wedge \, \mathrm{rsym}(c) = v \wedge \mathrm{state}(c) = s \wedge c \searrow^{\{\rightsquigarrow, \downarrow\}^*} f \Big) \Big\}.$$

For every $n \geq 0$ we set

$$L_n(A) := \left\{ v \in \Sigma^* \;\middle|\; \exists f \in F : \exists c \in \mathrm{mc}(A) : \exists x \in \{\rightsquigarrow, \downarrow\}^* : \Big(\mathrm{lsym}(c) = \lambda \right.$$
$$\left. \wedge \, \mathrm{rsym}(c) = v \wedge \mathrm{state}(c) = s \wedge |x|_\downarrow \leq n \wedge c \searrow^x f \Big) \right\}.$$

This gives us $L_R(A) = \cup_{n \geq 0} L_n(A)$.

Now let $c \in \mathrm{mc}(A)$, $q \in Q$, $i, k \in \mathbb{N}$ with $i \leq k$, $a_1, a_2, \ldots, a_k \in \Sigma$, and $n_1, n_2, \ldots, n_k \in \mathbb{N}_{>0}$ with $n_1 < n_2 < \cdots < n_k$ such that

$$c = (a_1, n_1)(a_2, n_2) \cdots (a_i, n_i) q (a_{i+1}, n_{i+1})(a_{i+2}, n_{i+2}) \cdots (a_k, n_k).$$

Furthermore, let $I \subseteq \mathrm{num}(c)$. We define the modified configuration $c \setminus I \in \mathrm{mc}(A)$ via $c \setminus I = \delta_1 \delta_2 \cdots \delta_i q \delta_{i+1} \delta_{i+2} \ldots \delta_k$, where for all $j \in \{1, 2, \ldots, k\}$ we set

$$\delta_j = \begin{cases} (a_j, n_j) & \text{if } n_j \notin I, \\ \lambda_{(\Sigma \times \mathbb{N}_{>0}) \cup Q} & \text{otherwise.} \end{cases}$$

Claim. Let $c, d \in \mathrm{mc}(A)$, $x \in \{\rightsquigarrow, \downarrow\}^*$ with $c \searrow^x d$, and $J \subseteq \mathrm{num}(d)$. Then, there is a sub-word y of x such that $(c \setminus J) \searrow^y (d \setminus J)$.

Proof. Consider the transformation from c to d that we get by using the symbols from x from left to right. We get y from x by deleting all symbols for which we jump over an element from $\Sigma \times J$ during this transformation. It should be clear that with this construction it holds $(c \setminus J) \searrow^y (d \setminus J)$. □

Now, all the notation we need for the proof of our theorem is established and we are ready for the main part of the proof.

Assume that there is a $u' \in \Sigma^*$ with $wu' \in L_R(A)$. Let $u \in \Sigma^*$ with $wu \in L_R(A)$ and

$$|u| = \min \left(\{ \, |u'| \mid u' \in \Sigma^* \wedge wu' \in L_R(A) \, \} \right).$$

Let $c \in \mathrm{mc}(A)$, $x \in \{\rightsquigarrow, \downarrow\}^*$, and $f \in F$ with $\mathrm{lsym}(c) = \lambda$, $\mathrm{rsym}(c) = wu$, $\mathrm{state}(c) = s$, $c \searrow^x f$, and $\mathrm{num}(c) = \{1, 2, \ldots, |wu|\}$. We need the following statement:

Claim. Let $d, e \in \mathrm{mc}(A)$ with

- $c \searrow^{\{\leadsto,\downarrow\}^*} d \searrow^{\{\leadsto,\downarrow\}^+} e \searrow^{\{\leadsto,\downarrow\}^*} f$,

- $\mathrm{lnum}(d) \subseteq \{1, 2, \ldots, |w|\} \supseteq \mathrm{lnum}(e)$, and

- $\mathrm{rnum}(d) \subseteq \{|w| + 1, |w| + 2, \ldots, |wu|\} \supseteq \mathrm{rnum}(e)$.

Then, we have $\mathrm{lnum}(e) \subset \mathrm{lnum}(d)$ or $\mathrm{state}(d) \neq \mathrm{state}(e)$.

Proof. It holds $\mathrm{lnum}(e) \subseteq \mathrm{lnum}(d)$ and $\mathrm{rnum}(e) \subseteq \mathrm{rnum}(d)$. Assume that $\mathrm{lnum}(d) = \mathrm{lnum}(e)$ and $\mathrm{state}(d) = \mathrm{state}(e)$. Then, there is a nonempty $I \subseteq \mathrm{rnum}(d)$ with $e = d \setminus I$. We get

$$(c \setminus I) \searrow^{\{\leadsto,\downarrow\}^*} (d \setminus I) = e \searrow^{\{\leadsto,\downarrow\}^*} f.$$

This implies $\mathrm{rsym}(c \setminus I) \in L_R(A)$, which is a contradiction because w is a prefix of $\mathrm{rsym}(c \setminus I)$ and $|\mathrm{rsym}(c \setminus I)| < |wu|$. So, it holds $\mathrm{lnum}(e) \subset \mathrm{lnum}(d)$ or $\mathrm{state}(d) \neq \mathrm{state}(e)$. $\square$

Set $R = \{\leadsto\}^{*\left((\{\leadsto,\downarrow\}^* \cdot \{\leadsto,\downarrow\})^\lambda \{\leadsto,\downarrow\}\right)}$. We can show the following:

Claim. It holds $wu \in L_{(|w|+1)\cdot|Q|-1}(A)$.

Proof. There exist modified configurations $c_0, c_1, \ldots, c_{|x|_\downarrow}$ such that

$$c \searrow^R c_0 \searrow^{R\downarrow R} c_1 \searrow^{R\downarrow R} c_2 \searrow^{R\downarrow R} \cdots \searrow^{R\downarrow R} c_{|x|_\downarrow} \searrow^R f$$

and for all $i \in \{0, 1, \ldots, |x|_\downarrow\}$ it holds $\mathrm{lnum}(c_i) \subseteq \{1, 2, \ldots, |w|\}$ and also $\mathrm{rnum}(c_i) \subseteq \{|w| + 1, |w| + 2, \ldots, |wu|\}$. For all $i, j \in \mathbb{N}$ with $i < j \leq |x|_\downarrow$ the just proven claim gives us $\mathrm{lnum}(c_j) \subset \mathrm{lnum}(c_i)$ or $\mathrm{state}(c_i) \neq \mathrm{state}(c_j)$. It follows $|x|_\downarrow < (|w| + 1) \cdot |Q|$, which implies that $wu \in L_{(|w|+1)\cdot|Q|-1}(A)$. $\square$

Let $n \geq 0$ and $\vec{\varphi} \in Q^n$. We define the DFA $A_{\vec{\varphi}}$ to be

$$\Big(Q^{n+1} \times \{0, 1\}, \Sigma, \delta_n, \{((s, \varphi_1, \varphi_2, \ldots, \varphi_n), 1)\},$$

$$\{((\varphi_1, \varphi_2, \ldots, \varphi_n, f), 1) \mid f \in F\}\Big),$$

where δ_n is defined as follows: Let $(q_1, q_2, \ldots, q_{n+1}) \in Q^{n+1}$, $k \in \{0,1\}$, and $a \in \Sigma$. If there is an $i \in \{1, 2, \ldots, n+1\}$ such that $a \in \Sigma_{\delta,q_i}$ and for all $j \in \{1, 2, \ldots, i-1\}$ it holds $a \notin \Sigma_{\delta,q_j}$, we set

$$\delta_n(((q_1, q_2, \ldots, q_{n+1}), k), a)$$
$$= ((q_1, q_2, \ldots, q_{i-1}, \delta(q_i, a), q_{i+1}, q_{i+2}, \ldots, q_{n+1}), k).$$

Otherwise, set

$$\delta_n(((q_1, q_2, \ldots, q_{n+1}), k), a) = ((q_1, q_2, \ldots, q_{n+1}), 0).$$

The DFA $A_{\vec{\varphi}}$ simultaneously simulates $n+1$ sweeps of A interpreted as a right one-way jumping finite automaton. The state vector $(s, \varphi_1, \varphi_2, \ldots, \varphi_n)$ contains the states that A is in at the beginning of the sweeps. The second component of a state of $A_{\vec{\varphi}}$ is 1 if and only if all yet considered symbols could be read in one of the $n+1$ sweeps. If a symbol cannot be read in one of the $n+1$ sweeps, the second component of the state of $A_{\vec{\varphi}}$ is set to 0, which means that the input word cannot be accepted by $A_{\vec{\varphi}}$.

The map δ_n is extended to words in the natural way. The following claim concerning δ_n can easily be seen *via* induction over $|v|$.

Claim. Let $v \in \Sigma^*$, $(q_1, q_2, \ldots, q_{n+1}) \in Q^{n+1}$, and $c_1 \in \mathrm{mc}(A)$ such that it is fulfilled $\mathrm{lsym}(c_1) = \lambda$, $\mathrm{rsym}(c_1) = v$, and $\mathrm{state}(c_1) = q_1$. Set

$$((p_1, p_2, \ldots, p_{n+1}), k) = \delta_n(((q_1, q_2, \ldots, q_{n+1}), 1), v).$$

Then, there are $c_2, c_3, \ldots, c_{n+1}, d_1, d_2, \ldots, d_{n+1} \in \mathrm{mc}(A)$ such that for all $i \in \{1, 2, \ldots, n+1\}$ and $j \in \{1, 2, \ldots, n\}$ it holds $c_i \searrow^R d_i$, $\mathrm{lsym}(c_i) = \lambda = \mathrm{rsym}(d_i)$, $\mathrm{rsym}(c_{j+1}) = \mathrm{lsym}(d_j)$, $\mathrm{rnum}(c_{j+1}) = \mathrm{lnum}(d_j)$, $\mathrm{state}(c_i) = q_i$, and $\mathrm{state}(d_i) = p_i$. Furthermore, we have $k = 1$ if and only if it holds $\mathrm{lsym}(d_{n+1}) = \lambda$. $\square$

From the previous claim, we directly get that $L(A_{\vec{\varphi}})$ equals

$$
\begin{aligned}
\Big\{ v \in \Sigma^* \ \Big| \ & \exists f \in F : \exists c \in \mathrm{mc}(A) : \exists m \in \{0, 1, \ldots, n\} : \\
& \exists d_1, d_2, \ldots, d_m \in \mathrm{mc}(A) : \\
& \Big(\mathrm{lsym}(c) = \lambda \wedge \mathrm{rsym}(c) = v \wedge \mathrm{state}(c) = s \\
& \wedge c \searrow^R d_1 \searrow^{\downarrow R} d_2 \searrow^{\downarrow R} \cdots \searrow^{\downarrow R} d_m \searrow^{\downarrow R} f \\
& \wedge \vec{\varphi} = (\mathrm{state}(d_1), \mathrm{state}(d_2), \ldots, \mathrm{state}(d_m), f, f, \ldots, f) \Big) \Big\}.
\end{aligned}
$$

This shows that for all $n \geq 0$ it holds $L_n(A) = \cup_{\vec{\varphi} \in Q^n} L(A_{\vec{\varphi}})$. Set $n = (|w| + 1) \cdot |Q| - 1$. From the third claim of this proof we know that there is a $\vec{\varphi} \in Q^n$ with $wu \in L(A_{\vec{\varphi}})$. For all $\vec{\psi} \in Q^{n+1}$ the state $(\vec{\psi}, 0)$ is a sink state of $A_{\vec{\varphi}}$. With the minimality of u, we get $|u| < |Q|^{n+1} = |Q|^{(|w|+1) \cdot |Q|}$. So, the condition that there is a $v \in L_R(A)$ such that w is a prefix of v is equivalent to the condition that there is a $u \in \Sigma^*$ with $|u| < |Q|^{(|w|+1) \cdot |Q|}$ and $wu \in L_R(A)$.

With analogous proofs one gets the following: The condition that there is a word $v \in L_R(A)$ such that w is a suffix of v is equivalent to the condition that there is a $u \in \Sigma^*$ with $|u| < |Q|^{(|w|+1) \cdot |Q|}$ and $uw \in L_R(A)$. The condition that there is a word $v \in L_R(A)$ such that w is a factor of v is equivalent to the condition that there are $u_1, u_2 \in \Sigma^*$ satisfying $|u_1| < |Q|^{(|w|+1) \cdot |Q|}$, $|u_2| < |Q|^{(|w|+1) \cdot |Q|}$, and $u_1 w u_2 \in L_R(A)$. Let $a_1, a_2, \ldots, a_{|w|} \in \Sigma$ with $w = a_1 a_2 \cdots a_{|w|}$. Then, the condition that there is a $v \in L_R(A)$ such that w is a sub-word of v is equivalent to the condition that there are words $u_0, u_1, \ldots, u_{|w|} \in \Sigma^*$ such that it holds $u_0 a_1 u_1 a_2 u_2 \cdots a_{|w|} u_{|w|} \in L_R(A)$ and for all $i \in \{0, 1, \ldots, |w|\}$ we have $|u_i| < |Q|^{(|w|+1) \cdot |Q|}$.

We prove that the first listed problem is in **PSPACE**, analogous proofs work for the other problems. The condition that there is a $v \in L_R(A)$ such that w is a prefix of v is equivalent to the condition that there is a $u \in \Sigma^*$ and a $\vec{\varphi} \in Q^n$ with $wu \in L(A_{\vec{\varphi}})$. A nondeterministic polynomial space bounded multi-tape Turing machine first guesses the state vector $\vec{\varphi}$. Then it simulates the computation of the DFA $A_{\vec{\varphi}}$ on the input word w. This can be done in polynomial bounded space, because the state set of $A_{\vec{\varphi}}$ is just $Q^{n+1} \times \{0, 1\}$. After that, u is guessed in a letter-by-letter fashion and the state of $A_{\vec{\varphi}}$ is updated accordingly. In this way it is verified if $wu \in L(A_{\vec{\varphi}})$. So, the problem if there is a $v \in L_R(A)$ such that w is a prefix of v, is in PSPACE. $\qquad\square$

7.3 ROWJFAs and Permutation Closed Languages

We consider the following problem for a fixed input alphabet. Given two NFAs A and B, do we have $L_J(A) \cap L_J(B) = \emptyset$? Fernau *et al.* stated that this problem is in P, referring to work by Eryk Kopczyński [45]. Using Corollary 5.5, we even get:

Corollary 7.6. *Let Σ be a fixed alphabet. Then, for two NFAs A and B with input alphabet Σ the problem to decide whether $L_J(A) \cap L_J(B) = \emptyset$, is NL-complete.*

Proof. The problem is NL-hard because already emptiness for NFAs over a fixed input alphabet is NL-complete. Let $k = |\Sigma|$ and $\Sigma = \{a_1, a_2, \ldots, a_k\}$. We show that the complement of the given problem is in NL. Let A and B be NFAs with input alphabet Σ and n be the maximum of the numbers of states of A and B. By Corollary 5.5, there exist a constant $c \in \mathbb{N}_{>0}$ and an NFA C such that C has at most $ck^{9k^2/2+43k/2+21}n^{6k^2+15k+15}$ states and $L_J(C) = L_J(A) \cap L_J(B)$. Hence, it holds $L_J(A) \cap L_J(B) \neq \emptyset$ if and only if there is a $w \in L_J(A) \cap L_J(B)$ with $|w| < ck^{9k^2/2+43k/2+21}n^{6k^2+15k+15}$. Our nondeterministic logspace bounded Turing machine T gets A and B as its input and guesses a $\vec{v} \in \mathbb{N}^k$ with $||\vec{v}||_\infty < ck^{9k^2/2+43k/2+21}n^{6k^2+15k+15}$. This vector is saved two times as $\vec{v_A}$ and $\vec{v_B}$. That can be done in space

$$
O\left(k \cdot \log\left(ck^{9k^2/2+43k/2+21}n^{6k^2+15k+15}\right)\right)
$$
$$
= O\big(k \cdot \log(c) + (9k^3/2 + 43k^2/2 + 21k)\log(k)
$$
$$
+ (6k^3 + 15k^2 + 15k)\log(n)\big).
$$

Then, T sets the current state of A to the initial state and repeats the following steps. An $i \in \{1, 2, \ldots, k\}$ is guessed such that $\pi_{k,i}(\vec{v_a}) > 0$ and a_i can be read in the current state of A. The current state of A is set to one of the possible successor states and the component i of $\vec{v_A}$ is decreased by one. These steps are repeated as often as possible. After that, the same is done for B and $\vec{v_B}$. The input is accepted by T if and only if $\vec{v_A}$ and $\vec{v_B}$ are both $\vec{0}$ at the end and A and B are both in an accepting state. □

Inclusion relations between the languages accepted by a given right one-way jumping finite automaton and a given ordinary jumping finite automaton are investigated. To do so, we need the following lemma.

Lemma 7.7. *Let A be a DFA with input alphabet Σ and n be the number of states of A. Then, a DFA B with $2n$ states fulfilling $L_J(B) = $ perm$(\Sigma^* \setminus L_R(A))$ is constructed by a deterministic logspace bounded Turing machine.*

Proof. Let $A = (Q, \Sigma, \delta, \{s\}, F)$ and set $Q' = \{q' \mid q \in Q\}$. The DFA B is defined as $(Q \cup Q', \Sigma, \gamma, \{s\}, (Q \setminus F) \cup Q')$, where the transition function γ is given as follows. For $(q, a) \in Q \times \Sigma$, it holds

$$\gamma(q, a) = \begin{cases} \delta(q, a) & \text{if } a \in \Sigma_{\delta, q}, \\ q' & \text{otherwise,} \end{cases} \qquad \gamma(q', a) = \begin{cases} \text{undefined} & \text{if } a \in \Sigma_{\delta, q}, \\ q' & \text{otherwise.} \end{cases}$$

It should be clear that B can be constructed by a deterministic logspace bounded Turing machine. We show that $L_J(B) = $ perm$(\Sigma^* \setminus L_R(A))$.

First, let $w \in L(B)$. There are two possibilities:

1. There is a $q \in Q \setminus F$ with $(s, w) \vdash_B^* (q, \lambda)$. Then we have $(s, w) \vdash_A^* (q, \lambda)$, which implies $w \notin L_R(A)$.

2. There is a $q \in Q$ with $(s, w) \vdash_B^* (q', \lambda)$. Then there is a $v \in (\Sigma \setminus \Sigma_{\delta, q})^* {(\Sigma^*, \Sigma, \lambda_\Sigma)} \setminus \{\lambda\}$ with $(s, w) \vdash_A^* (q, v)$. It follows $w \notin L_R(A)$.

That shows $L(B) \subseteq \Sigma^* \setminus L_R(A)$, which gives us $L_J(B) \subseteq $ perm$(\Sigma^* \setminus L_R(A))$.

Now, let $w \in \Sigma^* \setminus L_R(A)$. There are, again, two possibilities:

1. There is a $q \in Q \setminus F$ with $(s, w) \circlearrowright_A^* (q, \lambda)$. By Lemma 6.1, there is a permutation v of w with $(s, v) \vdash_A^* (q, \lambda)$. That implies $(s, v) \vdash_B^* (q, \lambda)$. Hence, we have $v \in L(B)$ and $w \in L_J(B)$.

2. There are a $q \in Q$ and a $v \in (\Sigma \setminus \Sigma_{\delta,q})^{*(\Sigma^*, \cdot \Sigma, \lambda_\Sigma)} \setminus \{\lambda\}$ with $(s, w) \circlearrowright_A^* (q, v)$. Lemma 6.1 tells us that there is an $x \in \Sigma^*$ such that xv is a permutation of w and $(s, x) \vdash_A^* (q, \lambda)$. It follows $(s, xv) \vdash_B^* (q, v) \vdash_B^* (q', \lambda)$, which implies $xv \in L(B)$ and $w \in L_J(B)$.

So, we have $\Sigma^* \setminus L_R(A) \subseteq L_J(B)$, which implies that it holds $L_J(B) = \mathsf{perm}\,(\Sigma^* \setminus L_R(A))$. □

We compare a language accepted by a right one-way jumping finite automaton with a language accepted by an ordinary jumping finite automaton. From Corollary 7.6, Lemma 7.7, and results by Kopczyński [45], we deduce:

Corollary 7.8. *Let A be a DFA and B be an NFA with the same input alphabet. Then, the problem to decide whether we have $L_R(A) \cap L_J(B) = \emptyset$ ($L_R(A) \subseteq L_J(B)$, $L_J(B) \subseteq L_R(A)$, respectively), is* NL-*complete (in* coNP, NL-*complete, respectively) if the input alphabet is fixed. If the alphabet is not fixed, the problem is in* coNP *(in* coNEXP, *in* coNP, *respectively).*

Proof. We have $L_R(A) \cap L_J(B) = \emptyset$ if and only if $L_J(A) \cap L_J(B) = \emptyset$. The inclusion $L_R(A) \subseteq L_J(B)$ holds if and only if $L_J(A) \subseteq L_J(B)$. By Lemma 7.7 a DFA C fulfilling $L_J(C) = \mathsf{perm}\,(\Sigma^* \setminus L_R(A))$ can be constructed by a deterministic logspace bounded Turing machine. We get $L_J(B) \subseteq L_R(A)$ if and only if $L_J(B) \cap L_J(C) = \emptyset$. The proposition follows from Corollary 7.6 and the results given by Kopczyński in the paper *Complexity of Problems of Commutative Grammars* in the table on page 25 [45]. □

For a DFA A the language $L_R(A)$ is closed under permutation if and only if $L_R(A) = L_J(A)$. Hence, Corollary 7.8 implies:

Corollary 7.9. *For a DFA A the problem whether $L_R(A)$ is permutation closed, is* NL-*complete if the input alphabet is fixed and of cardinality at least two. If the alphabet is not fixed, the problem is in* coNP. □

169

Proof. It only remains to show that for a DFA A the problem whether the language $L_R(A)$ is permutation closed, is **NL**-hard if the input alphabet is fixed and of cardinality at least two. We use a similar reduction as in the proof of Proposition 7.2. Let the graph G be defined as in this proof and let A be the DFA $(V \cup \{0, -1\}, \{a, b\}, \delta, \{0\}, \{m\})$, where

$$\delta = \{\, ib \to j, ia \to k \mid (i, j), (i, k) \in E \land j < k \,\} \cup \{0b \to 1, 0a \to -1\}.$$

If there is no path from 1 to m in G, we have $L_R(A) = \emptyset$. Otherwise, there is a $w \in \{a, b\}^*$ such that $bwa \in L_R(A)$, but $abw \notin L_R(A)$. □

If the input alphabet is fixed, the word problem for right one-way jumping finite automata that accept a permutation closed language is in **L**:

Proposition 7.10. *Let Σ be a fixed alphabet. Then, for a $w \in \Sigma^*$ and a DFA A with input alphabet Σ such that $L_R(A) = \mathsf{perm}(L_R(A))$ the problem to decide whether it holds $w \in L_R(A)$, is in* **L**.

Proof. In the paper *Characterization and Complexity Results on Jumping Finite Automata* by Fernau *et al.* the inclusion **JFA** $\subseteq$ **NST**$(\log(n), n)$ is shown at the beginning of Section 8 [29]. We only have to change one thing in their proof to show the proposition: In each step of the simulation the outgoing transition of the current state of the considered DFA M is not chosen nondeterministically, but deterministically in the following way. We order Σ arbitrarily and always choose the first input symbol for which there exists an outgoing transition of the current state and the corresponding counter is still positive. If there is no such input symbol, the input word is rejected. By modifying the algorithm in that manner, it decides whether $w \in L_R(M)$ because $L_R(M)$ can be assumed to be permutation closed in our case. □

For a fixed input alphabet the problems emptiness, finiteness, and universality of permutation closed languages accepted by right one-way jumping finite automata are all in **L**:

Proposition 7.11. *Let Σ be a fixed alphabet. Then, for a DFA A with input alphabet Σ such that $L_R(A) = \mathsf{perm}(L_R(A))$ the problems to decide* whether $L_R(A) = \emptyset$, whether $L_R(A) < \infty$, and whether $L_R(A) = \Sigma^*$, are all in L.

Proof. Let $A = (Q, \Sigma, \delta, \{s\}, F)$ be a DFA with $L_R(A) = \mathsf{perm}(L_R(A))$ and set $k = |\Sigma|$ and $n = |Q|$. We have $L_R(A) = \emptyset$ if and only if

$$L_R(A) \cap \psi^{-1}\left(\left\{ \vec{x} \in \mathbb{N}^k \ \middle| \ \sum_{i=1}^{k} x_i < n \right\}\right) = \emptyset.$$

It holds $L_R(A) = \Sigma^*$ if and only if

$$\psi^{-1}\left(\left\{ \vec{x} \in \mathbb{N}^k \ \middle| \ \sum_{i=1}^{k} x_i \le n \right\}\right) \subseteq L_R(A).$$

We get $L_R(A) < \infty$ if and only if

$$L_R(A) \cap \psi^{-1}\left(\left\{ \vec{x} \in \mathbb{N}^k \ \middle| \ n \le \sum_{i=1}^{k} x_i < 2n \right\}\right) = \emptyset.$$

All these properties can be checked in deterministic logarithmic space by Proposition 7.10. □

Using Corollary 5.5 we show that disjointness and inclusion for permutation closed languages accepted by right one-way jumping finite automata over a fixed input alphabet are in L:

Corollary 7.12. *Let Σ be a fixed alphabet. Then, for DFAs A and B with input alphabet Σ such that $L_R(A) = \mathsf{perm}(L_R(A))$ and $L_R(B) = \mathsf{perm}(L_R(B))$ the problems to decide whether $L_R(A) \cap L_R(B) = \emptyset$ and whether $L_R(A) \subseteq L_R(B)$, are in L.*

Proof. Let A and B be DFAs with input alphabet Σ such that $L_R(A) = \mathsf{perm}(L_R(A))$ and $L_R(B) = \mathsf{perm}(L_R(B))$. Let $k = |\Sigma|$ and n be the maximum of the numbers of states of A and B. By Corollary 5.5 there exist a constant $c \in \mathbb{N}_{>0}$ and an NFA C such that the automaton C has at most $ck^{9k^2/2+43k/2+21}n^{6k^2+15k+15}$ states and $L_J(C) = L_R(A) \cap L_R(B)$. Hence, it holds $L_R(A) \cap L_R(B) = \emptyset$ if and only if

$$L_R(A) \cap L_R(B) \cap \psi^{-1}\left(\left\{ \vec{x} \in \mathbb{N}^k \ \middle| \ \sum_{i=1}^{k} x_i < ck^{9k^2/2+43k/2+21}n^{6k^2+15k+15} \right\}\right)$$

is empty. This can be checked in deterministic logarithmic space by Proposition 7.10. Because of Lemma 7.7, there is DFA D with $2n$ states such that $L_J(D) = \Sigma^* \setminus L_R(B)$. It holds $L_R(A) \setminus L_R(B) = L_R(A) \cap L_J(D)$. By Corollary 5.5 there is an NFA E with at most $ck^{9k^2/2+43k/2+21}(2n)^{6k^2+15k+15}$ states and $L_J(E) = L_R(A) \cap L_J(D)$. Thus, we have $L_R(A) \subseteq L_R(B)$ if and only if $L_R(A) \setminus L_R(B)$ and

$$\psi^{-1}\left(\left\{ \vec{x} \in \mathbb{N}^k \;\middle|\; \sum_{i=1}^{k} x_i < ck^{9k^2/2+43k/2+21}(2n)^{6k^2+15k+15} \right\}\right)$$

are disjoint. □

We consider the problem to minimize the number of accepting states of a DFA interpreted as a right one-way jumping finite automaton that accepts a permutation closed language over a fixed input alphabet:

Corollary 7.13. *For every alphabet Σ there is a deterministic logarithmic space bounded Turing machine that computes for every DFA A with input alphabet Σ and $L_R(A) = \mathsf{perm}(L_R(A))$ a DFA B such that $L_R(B) = L_R(A)$ and the number of accepting states of B equals the cardinality of $L_R(B)/\sim_{L_R(B)}$.*

Proof. Let $A = (Q, \Sigma, \delta, \{s\}, F)$ be a DFA with $L_R(A) = \mathsf{perm}(L_R(A))$ such that every state is reachable from s. Thus, for all $p \in Q$ the language $L_R((Q, \Sigma, \delta, \{p\}, F))$ is permutation closed. The equivalence relation $=_R$ on Q is defined as follows. For all states $p, q \in Q$ we have $p =_R q$ if and only if

$$L_R((Q, \Sigma, \delta, \{p\}, F)) = L_R((Q, \Sigma, \delta, \{q\}, F)).$$

If Σ is fixed, for all $p, q \in Q$ it can be checked in deterministic logarithmic space if $p =_R q$ by Corollary 7.12. Let B be the DFA

$$(Q/=_R, \Sigma, \delta', \{[s]\}, \{\, [f] \mid f \in F \,\}),$$

where for all $(p, a) \in Q \times \Sigma$ for which there is a $q \in [p]$ such that $\delta(q, a)$ is defined we set $\delta'([p], a) = [\delta(q, a)]$ and for all $(p, a) \in Q \times \Sigma$ such that for all $q \in [p]$ the value $\delta(q, a)$ is undefined we let $\delta'([p], a)$ be undefined. This gives us $L_R(B) = L_R(A)$. For all states $p, q \in Q$ and words $v, w \in \Sigma^*$ with $v \sim_{L_R(A)} w$, $(s, v) \vdash_A^* (p, \lambda)$, and $(s, w) \vdash_A^* (q, \lambda)$ it holds $p =_R q$. It follows $|F/=_R| = |L_R(A)/\sim_{L_R(A)}|$. $\qquad \square$

Notice that by Corollaries 7.8 and 7.9 there is a polynomial space bounded Turing machine T that minimizes the number of states of a DFA A interpreted as a right one-way jumping finite automaton that accepts a permutation closed language. The machine T just checks for every DFA B whose number of states is smaller than the number of states of A whether $L_R(B) = L_R(A)$. For n-semirecognizability of a permutation closed language accepted by a right one-way jumping finite automaton, it holds:

Corollary 7.14. *For an $n \geq 0$ and a DFA A the problem to decide whether we have $L_R(A) \in$ **pROWJ$_n$**, is in* coNP.

Proof. To show that the complement of the given problem is in NP, we give a nondeterministic Turing-machine T that gets an $n \geq 0$ and a DFA $A = (Q, \Sigma, \delta, \{s\}, F)$ as its input. We can assume $L_R(A) = \mathsf{perm}(L_R(A))$ because of Corollary 7.9. Hence, for all $p \in Q$ that are reachable from s, the language $L_R((Q, \Sigma, \delta, \{p\}, F))$ is permutation closed. Let $m = |F|$ and $F = \{f_1, f_2, \ldots, f_m\}$. First, T initializes a counter c with the value 0. Then, for i from 1 to m the following is done. It is nondeterministically checked in logarithmic space whether f_i is reachable from s. If we get a positive answer, for all $j \in \{1, 2, \ldots, i-1\}$ we nondeterministically check if f_j is not reachable from s or

$$L_R((Q, \Sigma, \delta, \{f_i\}, F)) \neq L_R((Q, \Sigma, \delta, \{f_j\}, F)).$$

Corollary 7.8 implies that this can be done in polynomial time. If these checks deliver a positive answer for all $j \in \{1, 2, \ldots, i-1\}$, the counter c is incremented by 1. After this is done for i from 1 to m, the machine T accepts the input if and only if the value of c is at least $n + 1$. $\qquad \square$

For a right one-way jumping finite automaton that accepts a permutation closed language with a minimal number of accepting states, regularity is in NL and for a fixed input alphabet it is even in L:

Proposition 7.15. *For a DFA A with $L_R(A) = \mathsf{perm}(L_R(A))$ such that the number of accepting states of A equals the cardinality of the quotient $L_R(A)/{\sim}_{L_R(A)}$, the problem to decide whether $L_R(A)$ is regular, is in* NL. *If the input alphabet is fixed, the same problem is in* L.

Proof. Let $A = (Q, \Sigma, \delta, \{s\}, F)$ be a DFA with $L_R(A) = \mathsf{perm}(L_R(A))$ and $|F| = |L_R(A)/{\sim}_{L_R(A)}|$. By Proposition 4.47 the language $L_R(A)$ is regular if and only if for each $f \in F$ and every $a \in \Sigma$ for which there is a $w \in L((Q, \Sigma, \delta, \{f\}, \{f\}))$ with $|w|_a > 0$, there exists an $n > 0$ with $a^n \in L((Q, \Sigma, \delta, \{f\}, \{f\}))$. So, it can be checked if $L_R(A)$ is not regular in nondeterministic logarithmic space by guessing w in a letter-by-letter fashion. Let $\Sigma = \{a_1, a_2, \dots, a_{|\Sigma|}\}$. For all $f \in F$ and $i \in \{1, 2, \dots, |\Sigma|\}$ the condition that there is a $w \in L((Q, \Sigma, \delta, \{f\}, \{f\}))$ with $|w|_{a_i} > 0$ is equivalent to the condition that $L_R((Q, \Sigma, \delta, \{f\}, \{f\}))$ and

$$
\psi^{-1}\left(\left\{\vec{x} \in \mathbb{N}^{|\Sigma|} \;\middle|\; x_i > 0 \land \sum_{j=1}^{|\Sigma|} x_j \leq 2(|Q| - 1) + 1\right\}\right)
$$

are not disjoint. By Proposition 7.10 this can be checked in deterministic logarithmic space if Σ is fixed. □

With the results from Section 4.4, we get that for a right one-way jumping finite automaton which accepts a permutation closed language with a minimal number of accepting states, acceptance by different kinds of pushdown automata is in NP and for a fixed input alphabet the same problems are even in L:

Theorem 7.16. *For a DFA A with $L_R(A) = \mathsf{perm}(L_R(A))$ such that the number of accepting states of A equals the cardinality of $L_R(A)/{\sim}_{L_R(A)}$, the following problems are all in* NP. *If the input alphabet is fixed, these problems are all in* L.

- *Is $L_R(A)$ context free?*
- *Is $L_R(A)$ accepted by a counter automaton?*
- *Is $L_R(A)$ deterministic context free?*

- Is $L_R(A)$ accepted by a λ-free deterministic pushdown automaton?

- Is $L_R(A)$ accepted by a deterministic counter automaton?

- Is $L_R(A)$ accepted by a λ-free deterministic counter automaton?

Proof. Let $A = (Q, \Sigma, \delta, \{s\}, F)$ be a DFA with $L_R(A) = \mathsf{perm}(L_R(A))$ and $|F| = |L_R(A)/\sim_{L_R(A)}|$. Set $k = |\Sigma|$ and $n = |Q|$. For all $f \in F$ let

$$W_f = \{\, a \in \Sigma \mid \exists w \in L((Q, \Sigma, \delta, \{f\}, \{f\})) : |w|_a > 0 \,\}.$$

For all $f \in F$ the set W_f can be computed in polynomial time. If Σ is fixed, for all $f \in F$ the set W_f can be computed in logarithmic space, see the proof of Proposition 7.15. Let

$$G = \{\, f \in F \mid \exists a \in W_f : \forall m \in \mathbb{N}_{>0} : a^m \notin L((Q, \Sigma, \delta, \{f\}, \{f\})) \,\}.$$

A nondeterministic Turing machine T that gets A as its input guesses for each $f \in G$ a set $Y_f \subset \mathbb{N}^k$ with $|Y_f| = |W_f| - 1$ and $||Y_f||_\infty < n$. For all $f \in G$ it can be checked in deterministic polynomial time with the Bareiss algorithm whether Y_f is a linearly independent subset of $\mathbb{R}^k$ and $\psi^{-1}(Y_f) \subseteq L_R((Q, \Sigma, \delta, \{f\}, \{f\}))$. Only if this is true for all $f \in G$, the machine T accepts the input. By Proposition 4.51, this happens if and only if $L_R(A)$ is context free, and in this case $L_R(A)$ is accepted by a counter automaton. If T accepts the input, using the sets Y_f it can be checked in deterministic polynomial time with Cramer's rule and the Bareiss algorithm whether $L_R(A)$ is deterministic context free, whether $L_R(A)$ is accepted by a λ-free deterministic pushdown automaton, whether $L_R(A)$ is accepted by a deterministic counter automaton, and whether $L_R(A)$ is accepted by a λ-free deterministic counter automaton, by Propositions 4.55, 4.57, 4.59, and 4.62.

Let now Σ be fixed. For all $f \in G$ one can compute in logarithmic space a set $Z_f \subseteq \psi(L_R((Q, \Sigma, \delta, \{f\}, \{f\})))$ which is a linearly independent subset of $\mathbb{R}^k$ with $||Z_f||_\infty < n$ such that for all $Z \subseteq \psi(L_R((Q, \Sigma, \delta, \{f\}, \{f\})))$ which are linearly independent subsets of $\mathbb{R}^k$ with $Z_f \subseteq Z$ it holds $Z_f = Z$. By Proposition 4.51, the language $L_R(A)$ is context free if and only if for all $f \in G$ we have $|Z_f| = |W_f| - 1$. Assume that this is the case and define for all $f \in G$ the set S_f as $\psi(L_R((Q, \Sigma, \delta, \{f\}, \{f\})))$. Then, for all $f \in G$ the vector $\vec{v}_{S_f}$ can be computed and stored in logarithmic space. By Propositions 4.55, 4.57, 4.59, and 4.62, it can be checked in deterministic logarithmic space whether $L_R(A)$ is deterministic context free, whether $L_R(A)$ is accepted by a λ-free deterministic pushdown automaton, whether $L_R(A)$ is accepted by a deterministic counter automaton, and whether $L_R(A)$ is accepted by a λ-free deterministic counter automaton. $\qquad\square$

8 Nondeterministic Right One-Way Jumping Finite Automata

We interpret different types of nondeterministic finite automata as right one-way jumping finite automata and study the languages accepted by these devices. Independently Szilárd Zsolt Fazekas *et al.* investigated nondeterministic right one-way jumping finite automata [27]. They showed that each permutation closed semilinear language and each complement of a language from **ROWJ** is accepted by an MNFA interpreted as a right one-way jumping finite automaton. We start our investigation with MDFAs interpreted as right one-way jumping finite automata and show that the Parikh-image of each permutation closed language accepted by such a device is a quasi-lattice. This implies that each such language over a binary alphabet is a finite union of permutation closed semirecognizable languages. A characterization of permutation closed languages accepted by MDFAs interpreted as right one-way jumping finite automata is given. We show that the family of languages accepted by MDFAs interpreted as right one-way jumping finite automata and the family of languages accepted by NFAs interpreted as right one-way jumping finite automata are incomparable. It is proven that each permutation closed semilinear language is accepted by an NFA interpreted as a right one-way jumping finite automaton. We give three interpretations of λ-MNFAs as right one-way jumping finite automata. For each interpretation λ-MNFAs accept the same languages as λ-NFAs. For two of these interpretations λ-MNFAs accept the same languages as MNFAs. For the third interpretation more languages are accepted. However, all accepted languages are semilinear.

Right one-way jumping finite automata are compared to right-revolving finite automata as defined by Bensch *et al.* [11]. By definition, each language from **ROWJ** is accepted by a deterministic right-revolving finite automaton and, analogously, each language that is accepted by a λ-MNFA interpreted as a right one-way jumping finite automaton is accepted by a nondeterministic right-revolving finite automaton. Chigahara *et al.* gave a sketch of a proof that there is a language which is not in **ROWJ** but accepted by a deterministic right-revolving finite automaton [14]. We show that there is a non-semilinear language which is accepted by a deterministic right-revolving finite automaton. The mentioned language is even permutation closed. On the other hand, there is a permutation closed language that is accepted by an MDFA interpreted as a right one-way jumping finite automaton but not by a deterministic right-revolving finite automaton. We compare right one-way jumping finite automata to finite-state acceptors with translucent letters as defined by Nagy and Otto [58]. There is a language that is accepted by a deterministic finite-state acceptor with translucent letters but not by a λ-MNFA interpreted as a right one-way jumping finite automaton. Conversely, there is a language from **ROWJ** that is not accepted by a nondeterministic finite-state acceptor with translucent letters. However, each language from **pROWJ** is accepted by a deterministic finite-state acceptor with translucent letters. Nagy and Otto showed that all languages accepted by nondeterministic finite-state acceptors with translucent letters are semilinear and that each permutation closed semilinear language is accepted by a nondeterministic finite-state acceptor with translucent letters [58]. We prove that the family of permutation closed languages accepted by deterministic finite-state acceptors with translucent letters is incomparable to the family of permutation closed languages accepted by MDFAs interpreted as right one-way jumping finite automata. An open problem on deterministic finite-state acceptors with translucent letters from Nagy and Otto [58] is solved by us. We improve complexity results on finite-state acceptors with translucent letters from Nagy and Otto [58] and Kovács and Nagy [57] by showing that each language accepted by a deterministic (nondeterministic, respectively) finite-state acceptor with translucent letters is in $\mathbf{DST}(\log(n), n^2/\log(n))$ ($\mathbf{NST}(\log(n), n^2/\log(n))$, respectively) and for all $p \in \mathbb{Q}$ with $0 < p \leq 1$ is in $\mathbf{DST}(n^p, n^{2-p})$ ($\mathbf{NST}(n^p, n^{2-p})$, respectively). On the other hand, there is a language accepted by a deterministic finite-state acceptor with translucent letters which for all functions $f, g : \mathbb{N} \rightarrow \mathbb{N}$ with $f(n) \cdot g(n) \in o(n^2)$ is not in $\mathbf{NST}(f(n), g(n))$.

8.1 ROWJFAs with Multiple Initial States and Semirecognizability

For an MDFA A we define the binary relation $\circlearrowright_A$ on the set of configurations of A exactly the same way as for DFAs. For an MDFA $A = (Q, \Sigma, \delta, S, F)$ the language accepted by A interpreted as a right one-way jumping finite automaton is

$$L_R(A) = \left\{\, w \in \Sigma^* \mid \exists s \in S, \, f \in F : (s, w) \circlearrowright_A^* (f, \lambda_\Sigma) \,\right\}.$$

Thus, for each MDFA A it holds $L(A) \subseteq L_R(A) \subseteq L_J(A)$. The family of languages accepted by MDFAs interpreted as right one-way jumping finite automata is denoted by $\mathbf{MROWJ}$. This family consists exactly of the finite unions of languages from $\mathbf{ROWJ}$. Since each language in $\mathbf{MROWJ}$ contains a Parikh-equivalent regular sublanguage, the deterministic context-free language $\{\, a^n b^n \mid n \geq 0 \,\} \subset \{a, b\}^*$ is not in $\mathbf{MROWJ}$.

Example 8.1. The permutation closed language

$$L = \left\{\, w \in \{a, b\}^* \mid |w|_b = 0 \vee |w|_b = |w|_a \,\right\}$$

is not in $\mathbf{ROWJ}$ by Example 6.5. However, L is in $\mathbf{MROWJ}$ as the union of two languages from $\mathbf{ROWJ}$.

For permutation closed languages accepted by MDFAs interpreted as right one-way jumping finite automata, we get:

Theorem 8.2. *The family* $\mathbf{pMROWJ}$ *does* not *contain any language whose Parikh-image is a 0-anti-lattice.*

Proof. Assume that there are $k > 0$, $S \subseteq \mathbb{N}^k$, and an ordered alphabet Σ with $|\Sigma| = k$ such that S is a 0-anti-lattice but $\psi_\Sigma^{-1}(S)$ is in $\mathbf{MROWJ}$. So, there are $\vec{v}, \vec{w} \in \mathbb{N}^k$, where for all $j \in \{1, 2, \ldots, k\}$ it holds $\pi_{k, \{j\}}(\vec{v}) > 0$, such that $\mathsf{L}(\vec{0}, \{\vec{v}\}) \cap S = \emptyset$ and $\mathsf{L}(\vec{w}, \{\vec{v}, \vec{w}\}) \subseteq S$. Clearly we have $\vec{w} \neq \vec{0}$ and $\vec{v} \neq \vec{0}$. Since it holds that we have $\mathsf{L}(\vec{0}, \{\vec{v}\}) \cap \mathsf{L}(\vec{w}, \{\vec{v}, \vec{w}\}) = \emptyset$, there are no $\lambda, \mu \in \mathbb{N}_{>0}$ with $\lambda \vec{v} = \mu \vec{w}$, which implies that $\{\vec{v}, \vec{w}\}$ is a linearly independent subset of $\mathbb{R}^k$. Let $A = (Q, \Sigma, \delta, S', F)$ be an MDFA with $L_R(A) = \psi^{-1}(S)$ and $S' = \{s_1, s_2, \ldots, s_{|S'|}\}$. For $p, q \in Q$ consider the language $L_{p,q} = L((Q, \Sigma, \delta, \{p\}, \{q\}))$. For all $p, q \in Q$ there are a number $n_{p,q} \geq 0$, vectors $\overrightarrow{c_{p,q,1}}, \overrightarrow{c_{p,q,2}}, \ldots, \overrightarrow{c_{p,q,n_{p,q}}} \in \mathbb{N}^k$, and finite sets $P_{p,q,1}, P_{p,q,2}, \ldots, P_{p,q,n_{p,q}} \in 2^{\mathbb{N}^k \setminus \{\vec{0}\}}$ with $\psi(L_{p,q}) = \cup_{i=1}^{n_{p,q}} \mathsf{L}\left(\overrightarrow{c_{p,q,i}}, P_{p,q,i}\right)$. Set $n_0 = \max(\{\, n_{f,g} \mid f, g \in F\,\})$. For $p, q \in Q$ and a number $i \in \{1, 2, \ldots, n_{p,q}\}$ let $P_{p,q,i} = \left\{\overrightarrow{x_{p,q,i,1}}, \overrightarrow{x_{p,q,i,2}}, \ldots, \overrightarrow{x_{p,q,i,|P_{p,q,i}|}}\right\}$ and the function $\kappa_{p,q,i} : \mathbb{N}^{|P_{p,q,i}|} \to \mathbb{N}^k$ be defined through

$$\left(\lambda_1, \lambda_2, \ldots, \lambda_{|P_{p,q,i}|}\right) \mapsto \sum_{j=1}^{|P_{p,q,i}|} \lambda_j \overrightarrow{x_{p,q,i,j}}.$$

For $f, g \in F$, $i \in \{1, 2, \ldots, n_{f,g}\}$, and $R \in \mathbb{Z}$ let

$$A_{f,g,i,R} = \left\{ \vec{\lambda} \in \mathbb{N}^{|P_{f,g,i}|} \;\middle|\; \overrightarrow{c_{f,g,i}} + \kappa_{f,g,i}\left(\vec{\lambda}\right) \in \mathsf{L}\left(\vec{0}, \{\vec{v}\}\right) + R\vec{w} \right\}.$$

Let X be the set

$$\left\{ (f, g, i) \in F \times F \times \mathbb{N} \;\middle|\; 1 \leq i \leq n_{f,g} \wedge \forall r \in \mathbb{N} : \bigcup_{R \in \mathbb{Z}, |R| \geq r} A_{f,g,i,R} \neq \emptyset \right\}$$

and let r_0 be the maximum of

$$\{0\} \cup \{|R| \;\mid\; R \in \mathbb{Z} \wedge \exists (f, g, i) \in (F \times F \times \mathbb{N}) \setminus X :$$
$$(1 \leq i \leq n_{f,g} \wedge A_{f,g,i,R} \neq \emptyset)\}.$$

For $f, g \in F$ and $i \in \{1, 2, \ldots, n_{f,g}\}$ let $D_{f,g,i} = \min\left(\cup_{R \in \mathbb{Z}} A_{f,g,i,R}\right)$ and $r_{f,g,i}$ be the maximum of

$$\{0\} \cup \left\{|R| \;\middle|\; R \in \mathbb{Z} \wedge \exists \vec{\lambda} \in D_{f,g,i} : \overrightarrow{c_{f,g,i}} + \kappa_{f,g,i}\left(\vec{\lambda}\right) \in \mathsf{L}\left(\vec{0}, \{\vec{v}\}\right) + R\vec{w} \right\}.$$

Let $(f, g, i) \in X$. There is an $R \in \mathbb{Z}$ with $|R| > r_{f,g,i}$ and $A_{f,g,i,R} \neq \emptyset$. Let $\vec{\lambda} \in A_{f,g,i,R}$. There are $\vec{\mu} \in D_{f,g,i}$ and $R' \in \mathbb{Z}$ with $\vec{\mu} \leq \vec{\lambda}$ and $\vec{\mu} \in A_{f,g,i,R'}$. It follows $|R'| \leq r_{f,g,i} < |R|$ and that there exists a $T \in \mathbb{Z}$ with

$$\kappa_{f,g,i}\left(\vec{\lambda} - \vec{\mu}\right) = \left(\overrightarrow{c_{f,g,i}} + \kappa_{f,g,i}\left(\vec{\lambda}\right)\right) - \left(\overrightarrow{c_{f,g,i}} + \kappa_{f,g,i}(\vec{\mu})\right) = T\vec{v} + (R - R')\vec{w}.$$

So, for all $(f, g, i) \in X$ there are $\overrightarrow{\lambda_{f,g,i}} \in \mathbb{N}^{|P_{f,g,i}|}$, $T_{f,g,i} \in \mathbb{Z}$, and $R_{f,g,i} \in \mathbb{Z} \setminus \{0\}$ with $\kappa_{f,g,i}\left(\overrightarrow{\lambda_{f,g,i}}\right) = T_{f,g,i} \cdot \vec{v} + R_{f,g,i} \cdot \vec{w}$. Let r_1 be the maximum of $\{1\} \cup \{\, |R_{f,g,i}| \mid (f, g, i) \in X \,\}$ and t_0 be the maximum of $\{0\} \cup \{\, |T_{f,g,i}| \mid (f, g, i) \in X \,\}$.

For $m \in \{0, 1, \ldots, |S'|\}$ we define the words $\alpha_m \in \Sigma^*$ recursively in the following manner. Set $\alpha_0 = \lambda$. Let now $m \in \{1, 2, \ldots, |S'|\}$, assume that the word α_{m-1} is already defined, and let $p_m \in Q$ and $\beta_m \in \Sigma^*$ such that $(s_m, \alpha_{m-1}) \circlearrowleft^{|\alpha_{m-1}|} (p_m, \beta_m)$. If

$$\bigcup_{f \in F} \left(\psi(L_{p_m, f}) \cap (\mathsf{L}(\vec{w}, \{\vec{v}, \vec{w}\}) - \{\psi(\alpha_{m-1}) + \psi(\beta_m)\}) \right) = \emptyset,$$

we set $\alpha_m = \alpha_{m-1}$. Otherwise, let $f_m \in F$ with

$$\psi(L_{p_m, f_m}) \cap (\mathsf{L}(\vec{w}, \{\vec{v}, \vec{w}\}) - \{\psi(\alpha_{m-1}) + \psi(\beta_m)\}) \neq \emptyset$$

and let

$$\gamma_m \in L_{p_m, f_m} \cap \psi^{-1}\left((\mathsf{L}(\vec{w}, \{\vec{v}, \vec{w}\}) - \{\psi(\alpha_{m-1}) + \psi(\beta_m)\}) \cap \mathbb{N}^k \right).$$

It follows $(s_m, \alpha_{m-1}\gamma_m) \circlearrowleft^{|\alpha_{m-1}\gamma_m|} (f_m, \beta_m)$. If there are no $g \in F$ and $i \in \mathbb{N}$ such that $(f_m, g, i) \in X$, we set $\alpha_m = \alpha_{m-1}\gamma_m$. Otherwise, let $g_m \in F$ and $i_m \in \mathbb{N}$ with $(f_m, g_m, i_m) \in X$. Furthermore, let $\delta_m \in L_{f_m, g_m}$ such that

$$\psi(\delta_m) = \overrightarrow{c_{f_m, g_m, i_m}} + |S'| \cdot |F| \cdot n_0 \cdot (2r_0 + 1) \cdot \frac{r_1!}{|R_{f_m, g_m, i_m}|} \cdot \kappa_{f_m, g_m, i_m}\left(\overrightarrow{\lambda_{f_m, g_m, i_m}}\right)$$

and set $\alpha_m = \alpha_{m-1}\gamma_m\delta_m$. It follows $(s_m, \alpha_m) \circlearrowleft^{|\alpha_m|} (g_m, \beta_m)$. This completes the recursive definition of the words α_m. Since $\vec{v} \in (\mathbb{N}_{>0})^k$, there is a $\zeta \in \Sigma^*$ such that $\psi(\alpha_{|S'|}\zeta) \in L(\vec{0}, \{\vec{v}\})$. Let $\eta \in \psi^{-1}(\{\vec{w}\})$ and for all $m \in \{0, 1, \ldots, |S'|\}$ let $\alpha_m' \in \Sigma^*$ such that it holds $\alpha_m \alpha_m' = \alpha_{|S'|}$.

Let now $m \in \{1, 2, \ldots, |S'|\}$ with

$$\bigcup_{f \in F} \left(\psi(L_{p_m, f}) \cap (\mathsf{L}(\vec{w}, \{\vec{v}, \vec{w}\}) - \{\psi(\alpha_{m-1}) + \psi(\beta_m)\}) \right) = \emptyset$$

and $K > 0$. It holds

$$\psi\left(\alpha'_m \zeta \eta^{K \cdot r_1!} \beta_m\right) = \psi\left(\alpha_{|S'|} \zeta \eta^{K \cdot r_1!}\right) - \psi(\alpha_{m-1}) + \psi(\beta_m)$$
$$\in \mathsf{L}(\vec{w}, \{\vec{v}, \vec{w}\}) - \{\psi(\alpha_{m-1}) + \psi(\beta_m)\}.$$

We have

$$\left(s_m, \alpha_{|S'|} \zeta \eta^{K \cdot r_1!}\right) = \left(s_m, \alpha_{m-1} \alpha'_m \zeta \eta^{K \cdot r_1!}\right) \circlearrowleft^{|\alpha_{m-1}|} \left(p_m, \alpha'_m \zeta \eta^{K \cdot r_1!} \beta_m\right).$$

For all $f \in F$ it holds $\psi(\alpha'_m \zeta \eta^{K \cdot r_1!} \beta_m) \notin \psi(L_{p_m, f})$. With an analogous argument as in the proof of Lemma 6.1 it follows $\alpha_{|S'|} \zeta \eta^{K \cdot r_1!} \notin L_R((Q, \Sigma, \delta, \{s_m\}, F))$.

Now, let $m \in \{1, 2, \ldots, |S'|\}$ such that

$$\bigcup_{f \in F} (\psi(L_{p_m, f}) \cap (\mathsf{L}(\vec{w}, \{\vec{v}, \vec{w}\}) - \{\psi(\alpha_{m-1}) + \psi(\beta_m)\})) \neq \emptyset$$

and there are no $g \in F$ and $i \in \mathbb{N}$ with $(f_m, g, i) \in X$. For all $g \in F$, $i \in \{1, 2, \ldots, n_{f_m, g}\}$, and $R \in \mathbb{Z}$ with $|R| > r_0$ we have $A_{f_m, g, i, R} = \emptyset$. For all $K > 0$ it holds

$$\left(s_m, \alpha_{|S'|} \zeta \eta^{K \cdot r_1!}\right) = \left(s_m, \alpha_{m-1} \gamma_m \alpha'_m \zeta \eta^{K \cdot r_1!}\right) \circlearrowleft^{|\alpha_{m-1} \gamma_m|} \left(f_m, \alpha'_m \zeta \eta^{K \cdot r_1!} \beta_m\right)$$

and

$$\psi\left(\alpha'_m \zeta \eta^{K \cdot r_1!} \beta_m\right) = \psi\left(\alpha_{|S'|} \zeta \eta^{K \cdot r_1!}\right) - (\psi(\gamma_m) + \psi(\alpha_{m-1}) - \psi(\beta_m))$$
$$= \psi\left(\alpha_{|S'|} \zeta\right) - (\psi(\gamma_m) + \psi(\alpha_{m-1}) - \psi(\beta_m)) + K \cdot r_1! \cdot \vec{w},$$

where $\psi(\alpha_{|S'|} \zeta) \in \mathsf{L}(\vec{0}, \{\vec{v}\})$ and $\psi(\gamma_m) + \psi(\alpha_{m-1}) - \psi(\beta_m) \in \mathsf{L}(\vec{w}, \{\vec{v}, \vec{w}\})$. If there is a $K > 0$ such that $\alpha_{|S'|} \zeta \eta^{K \cdot r_1!} \in L_R((Q, \Sigma, \delta, \{s_m\}, F))$, we have

$$\psi\left(\alpha'_m \zeta \eta^{K \cdot r_1!} \beta_m\right) \in \bigcup_{g \in F} \bigcup_{i=1}^{n_{f_m, g}} \mathsf{L}(\overrightarrow{c_{f_m, g, i}}, P_{f_m, g, i}).$$

It follows

$$\left| \left\{ \alpha_{|S'|} \zeta \eta^{K \cdot r_1!} \,\middle|\, K > 0 \right\} \cap L_R((Q, \Sigma, \delta, \{s_m\}, F)) \right| \leq |F| \cdot n_0 \cdot (2r_0 + 1).$$

Because for all $K > 0$ it holds $\psi(\alpha_{|S'|}\zeta\eta^{K \cdot r_1!}) \in \mathsf{L}(\vec{w}, \{\vec{v}, \vec{w}\}) \subseteq S$, there are a $K > 0$ with

$$K \le (|S'| - 1) \cdot |F| \cdot n_0 \cdot (2r_0 + 1) + 1 \le |S'| \cdot |F| \cdot n_0 \cdot (2r_0 + 1)$$

and an $m \in \{1, 2, \ldots, |S'|\}$ such that

$$\bigcup_{f \in F} \left(\psi(L_{p_m, f}) \cap (\mathsf{L}(\vec{w}, \{\vec{v}, \vec{w}\}) - \{\psi(\alpha_{m-1}) + \psi(\beta_m)\}) \right) \ne \emptyset$$

and also $\alpha_m \ne \alpha_{m-1}\gamma_m$ and $\alpha_{|S'|}\zeta\eta^{K \cdot r_1!} \in L_R((Q, \Sigma, \delta, \{s_m\}, F))$. We have

$$\left(s_m, \alpha_{|S'|}\zeta\eta^{K \cdot r_1!} \right) \circlearrowleft^{|\alpha_m|} \left(g_m, \alpha_m'\zeta\eta^{K \cdot r_1!}\beta_m \right).$$

Let $\delta_m' \in L_{f_m, g_m}$ such that $\psi(\delta_m')$ equals

$$\overrightarrow{c_{f_m, g_m, i_m}} + \left(|S'| \cdot |F| \cdot n_0 \cdot (2r_0 + 1) - \frac{R_{f_m, g_m, i_m}}{|R_{f_m, g_m, i_m}|} \cdot K \right)$$
$$\cdot \frac{r_1!}{|R_{f_m, g_m, i_m}|} \cdot \kappa_{f_m, g_m, i_m} \left(\overrightarrow{\lambda_{f_m, g_m, i_m}} \right).$$

We get

$$\left(s_m, \alpha_{m-1}\gamma_m\delta_m'\alpha_m'\zeta\eta^{K \cdot r_1!} \right) \circlearrowleft^{|\alpha_{m-1}\gamma_m\delta_m'|} \left(g_m, \alpha_m'\zeta\eta^{K \cdot r_1!}\beta_m \right),$$

which implies

$$\alpha_{m-1}\gamma_m\delta_m'\alpha_m'\zeta\eta^{K \cdot r_1!} \in L_R((Q, \Sigma, \delta, \{s_m\}, F)) \subseteq \psi^{-1}(S).$$

It holds

$$\psi\left(\alpha_{m-1}\gamma_m\delta_m'\alpha_m'\zeta\eta^{K \cdot r_1!} \right) - \psi\left(\alpha_{|S'|}\zeta\eta^{K \cdot r_1!} \right)$$
$$= \psi(\delta_m') - \psi(\delta_m)$$
$$= -\frac{R_{f_m, g_m, i_m}}{|R_{f_m, g_m, i_m}|} \cdot K \cdot \frac{r_1!}{|R_{f_m, g_m, i_m}|} \cdot \kappa_{f_m, g_m, i_m} \left(\overrightarrow{\lambda_{f_m, g_m, i_m}} \right)$$
$$= \frac{-K \cdot r_1!}{R_{f_m, g_m, i_m}} \cdot (T_{f_m, g_m, i_m} \cdot \vec{v} + R_{f_m, g_m, i_m} \cdot \vec{w})$$
$$= \frac{-K \cdot r_1! \cdot T_{f_m, g_m, i_m}}{R_{f_m, g_m, i_m}} \cdot \vec{v} - K \cdot r_1! \cdot \vec{w}.$$

Because of $\psi(\alpha_{|S'|}\zeta\eta^{K \cdot r_1!}) \in \mathsf{L}\,(K \cdot r_1! \cdot \vec{w}, \{\vec{v}\})$, we get

$$\psi\left(\alpha_{m-1}\gamma_m\delta'_m\alpha'_m\zeta\eta^{K \cdot r_1!}\right) \in \mathsf{L}\left(\vec{0}, \{\vec{v}\}\right),$$

which gives us $\psi(\alpha_{m-1}\gamma_m\delta'_m\alpha'_m\zeta\eta^{K \cdot r_1!}) \notin S$, a contradiction. □

Example 8.3. The Parikh-image of the permutation closed deterministic context-free language $L = \{\,w \in \{a,b\}^* \mid |w|_a \neq |w|_b\,\}$ was shown to be a 0-anti-lattice in Example 4.37. Thus, L is not in **MROWJ**.

To extend Theorem 8.2 to arbitrary anti-lattices, we use the following:

Lemma 8.4. *Let Σ be an alphabet, $a \in \Sigma$, and $L \subseteq \Sigma^*$ be in the class pMROWJ. Then, the language $\{\,vaw \mid v, w \in \Sigma^* \land vw \in L\,\}$ is also in pMROWJ.*

Proof. Let $A = (Q, \Sigma, \delta, S, F)$ be an MDFA with $L_R(A) = L$. Then, we consider the MDFA $A' = (Q \cup S', \Sigma, \delta', S', F)$, where $S' = \{\,s' \mid s \in S\,\}$ and for all $(q, b) \in Q \times \Sigma$ the value $\delta'(q, b)$ is defined if and only if $\delta(q, b)$ is defined. In this case we have that it holds $\delta'(q, b) = \delta(q, b)$. For $s \in S$ we get $\delta'(s', a) = s$ and for all $b \in \Sigma \setminus \{a\}$ the value $\delta'(s', b)$ is undefined. Obviously, for every $w \in L_R(A')$ we have $|w|_a > 0$. For $v, w \in \Sigma^*$ with $|v|_a = 0$ and $s \in S$ it holds $(s', vaw) \circlearrowright_{A'}^{|v|} (s', awv) \circlearrowright_{A'} (s, wv)$. This gives us

$$vaw \in L_R(A') \Leftrightarrow wv \in L_R(A) \Leftrightarrow vw \in L_R(A),$$

which implies

$$\begin{aligned} L_R(A') &= \{\,vaw \mid v, w \in \Sigma^* \land |v|_a = 0 \land vw \in L\,\} \\ &= \{\,vaw \mid v, w \in \Sigma^* \land vw \in L\,\}. \end{aligned}$$

□

Theorem 8.2 and Lemmas 4.38 and 8.4 imply:

Corollary 8.5. *The family pMROWJ does not contain any language whose Parikh-image is an anti-lattice.* □

With Proposition 4.39 we get:

Corollary 8.6. *The Parikh-image of each language in* **pMROWJ** *is a quasi-lattice.* □

Proposition 4.41 gives us:

Corollary 8.7. *A language over a binary alphabet is in* **pMROWJ** *if and only if it is a finite union of permutation closed semirecognizable languages.* □

To get a characterization result about **pMROWJ** we use the following notion. For an alphabet Σ, $w \in \Sigma^*$, and $L \subseteq \Sigma^*$ the *disjoint quotient* of L with w is

$$L/^d w = \left\{ v \in \Sigma^* \mid vw \in L \wedge \forall a \in \Sigma : |v|_a \cdot |w|_a = 0 \right\}.$$

By Theorem 6.4 the family **pROWJ** is closed under the operation of disjoint quotient with a word. Corollary 8.6 and Proposition 4.40 imply:

Corollary 8.8. *Let Σ be an alphabet and $L \subseteq \Sigma^*$. Then, L is a finite union of permutation closed semirecognizable languages if and only if for all $w \in \Sigma^*$ it holds that $L/^d w \in$* **pMROWJ**. □

Thus, every language in **pMROWJ** is a finite union of permutation closed semirecognizable languages if and only if **pMROWJ** is closed under the operation of disjoint quotient with a word. It is an open problem if **pMROWJ** is closed under disjoint quotient with a word.

8.2 Nondeterministic ROWJFAs

For an MNFA A the binary relation $\circlearrowright_A$ on the set of configurations of A and the language $L_R(A)$ are defined exactly as for MDFAs. The family of languages accepted by MNFAs (NFAs, respectively) interpreted as right one-way jumping finite automata is denoted by **MNROWJ** (**NROWJ**, respectively). Independently from us, Fazekas *et al.* showed that each permutation closed semilinear language is in **MNROWJ** [27]. We get:

Proposition 8.9. *Each permutation closed semilinear language language is in* **NROWJ**.

Proof. Let $k \geq 1$, $n \geq 0$, $\vec{c}_1, \vec{c}_2, \ldots, \vec{c}_n \in \mathbb{N}^k$, and $P_1, P_2, \ldots, P_n$ be finite subsets of $\mathbb{N}^k \setminus \{\vec{0}\}$. We set

$$Q = \{s\} \cup \left\{ p_{i,\vec{x}},\, q_{i,\vec{x}} \;\middle|\; i \in \{1, 2, \ldots, n\} \wedge \vec{x} \in \mathbb{N}^k \wedge \vec{x} \leq \vec{c}_i \right\}$$

$$\cup \left\{ r_{i,\vec{y},\vec{x}},\, t_{i,\vec{y},\vec{x}} \;\middle|\; i \in \{1, 2, \ldots, n\} \wedge \vec{y} \in P_i \wedge \vec{x} \in \mathbb{N}^k \wedge \vec{x} \leq \vec{y} \right\}.$$

Let $\simeq$ be the equivalence relation on Q that is generated by

$$\left\{ \left(s, q_{i,\vec{0}} \right) \;\middle|\; i \in \{1, 2, \ldots, n\} \right\}$$

$$\cup \left\{ \left(s, r_{i,\vec{y},\vec{0}} \right) \;\middle|\; i \in \{1, 2, \ldots, n\} \wedge \vec{y} \in P_i \right\}$$

$$\cup \left\{ \left(r_{i,\vec{y},\vec{y}}, p_{i,\vec{0}} \right) \;\middle|\; i \in \{1, 2, \ldots, n\} \wedge \vec{y} \in P_i \right\}$$

$$\cup \left\{ \left(p_{i,\vec{x}}, q_{i,\vec{x}} \right) \;\middle|\; i \in \{1, 2, \ldots, n\} \wedge \vec{x} \in \mathbb{N}^k \setminus \{\vec{0}\} \wedge \vec{x} \leq \vec{c}_i \right\}$$

$$\cup \left\{ \left(p_{i,\vec{c}_i}, t_{i,\vec{y},\vec{0}} \right) \;\middle|\; i \in \{1, 2, \ldots, n\} \wedge \vec{y} \in P_i \right\}$$

$$\cup \left\{ \left(t_{i,\vec{y},\vec{y}}, p_{i,\vec{c}_i} \right) \;\middle|\; i \in \{1, 2, \ldots, n\} \wedge \vec{y} \in P_i \right\}.$$

Let Σ be an alphabet with $|\Sigma| = k$ and A be the NFA $(Q/{\simeq}, \Sigma, \delta, \{[s]_{\simeq}\}, F)$, where

$$F = \left\{ [p_{i,\vec{c}_i}]_{\simeq}, [q_{i,\vec{c}_i}]_{\simeq} \;\middle|\; i \in \{1, 2, \ldots, n\} \right\}$$

and δ is

$$\left\{ [u_{i,\vec{x}}]_{\simeq} a \to [u_{i,\vec{x}+\psi(a)}]_{\simeq} \quad\middle|\; a \in \Sigma \wedge u \in \{p, q\} \wedge i \in \{1, 2, \ldots, n\} \right.$$

$$\left. \wedge \vec{x} \in \mathbb{N}^k \wedge \vec{x} + \psi(a) \leq \vec{c}_i \right\}$$

$$\cup \left\{ [u_{i,\vec{y},\vec{x}}]_{\simeq} a \to [u_{i,\vec{y},\vec{x}+\psi(a)}]_{\simeq} \quad\middle|\; a \in \Sigma \wedge u \in \{r, t\} \wedge i \in \{1, 2, \ldots, n\} \right.$$

$$\left. \wedge \vec{y} \in P_i \wedge \vec{x} \in \mathbb{N}^k \wedge \vec{x} + \psi(a) \leq \vec{y} \right\}.$$

We show that $L_R(A) = \psi^{-1}\left(\bigcup_{i=1}^{n} \mathsf{L}(\vec{c}_i, P_i) \right)$. First, let $w \in L(A)$. From our definitions one can see that there are $i \in \{1, 2, \ldots, n\}$, $w_1 \in \psi^{-1}(P_i) \cup \{\lambda\}$, $w_2 \in \psi^{-1}(\{\vec{c}_i\})$, and $w_3 \in \left(\psi^{-1}(P_i) \right)^{*(\Sigma^*, \cdot_\Sigma, \lambda_\Sigma)}$ with $w = w_1 w_2 w_3$. It follows $w \in \psi^{-1}\left(\bigcup_{i=1}^{n} \mathsf{L}(\vec{c}_i, P_i)\right)$. This gives us $L_R(A) \subseteq \psi^{-1}\left(\bigcup_{i=1}^{n} \mathsf{L}(\vec{c}_i, P_i)\right)$. To show $\psi^{-1}\left(\bigcup_{i=1}^{n} \mathsf{L}(\vec{c}_i, P_i)\right) \subseteq L_R(A)$, we prove the following claim *via* induction over m.

Claim. Let $a \in \Sigma$, $w \in \Sigma^*$, and $i \in \{1, 2, \ldots, n\}$ with $\psi(a) \le \vec{c}_i \le \psi(aw)$. Furthermore, let $m \in \left\{0, 1, \ldots, \Sigma_{j=1}^{k} \pi_{k,j}(\vec{c}_i)\right\}$. Then, there are $v \in \Sigma^*$ and $\vec{x} \in \mathbb{N}^k$ with $\Sigma_{j=1}^{k} x_j = m$ and $\vec{x} \le \vec{c}_i$ such that $\vec{x} + \psi(v) = \psi(aw)$ and $([s]_\simeq, aw) \circlearrowright^m ([q_{i,\vec{x}}]_\simeq, v)$.

Proof. For $m = 0$ we just set $v = aw$ and $\vec{x} = \vec{0}$. For $m = 1$ we have

$$([s]_\simeq, aw) = \left(\left[q_{i,\vec{0}}\right]_\simeq, aw\right) \circlearrowright \left(\left[q_{i,\psi(a)}\right]_\simeq, w\right),$$

so the claim is fulfilled in this case. Let now $m \in \left\{2, 3, \ldots, \Sigma_{j=1}^{k} \pi_{k,j}(\vec{c}_i)\right\}$. By the induction hypothesis there are $v \in \Sigma^*$ and $\vec{x} \in \mathbb{N}^k$ with $\Sigma_{j=1}^{k} x_j = m - 1$ such that $\vec{x} \le \vec{c}_i \le \psi(aw) = \vec{x} + \psi(v)$ and $([s]_\simeq, aw) \circlearrowright^{m-1} ([q_{i,\vec{x}}]_\simeq, v)$. It follows that the inequality $\Sigma_{j=1}^{k} x_j < \Sigma_{j=1}^{k} \pi_{k,j}(\vec{c}_i)$ holds. For each $b \in \Sigma$ we have $b \in \Sigma_{\delta,[q_{i,\vec{x}}]_\simeq}$ if and only if $\vec{x} + \psi(b) \le \vec{c}_i$. So, there are $c \in \Sigma_{\delta,[q_{i,\vec{x}}]_\simeq}$, $v' \in \left(\Sigma \setminus \Sigma_{\delta,[q_{i,\vec{x}}]_\simeq}\right)^{*(\Sigma^*, \Sigma, \lambda_\Sigma)}$, and $v'' \in \Sigma^*$ such that $v = v'cv''$. We get $([q_{i,\vec{x}}]_\simeq, v) \circlearrowright ([q_{i,\vec{x}+\psi(c)}]_\simeq, v''v')$ and

$$\vec{x} + \psi(c) + \psi(v''v') = \vec{x} + \psi(v) = \psi(aw).$$

$\square$

If we set $m = \Sigma_{j=1}^{k} \pi_{k,j}(\vec{c}_i)$ in the claim, we get the first of the following statements. The other three statements can be proved analogously.

- For all $a \in \Sigma$, $w \in \Sigma^*$, and $i \in \{1, 2, \ldots, n\}$ with $\psi(a) \le \vec{c}_i \le \psi(aw)$, there is a $v \in \Sigma^*$ such that $\vec{c}_i + \psi(v) = \psi(aw)$ and $([s]_\simeq, aw) \circlearrowright^* ([q_{i,\vec{c}_i}]_\simeq, v)$.

- For all $a \in \Sigma$, $w \in \Sigma^*$, $i \in \{1, 2, \ldots, n\}$, and $\vec{y} \in P_i$ with $\psi(a) \le \vec{y} \le \psi(aw)$, there is a $v \in \Sigma^*$ such that $\vec{y} + \psi(v) = \psi(aw)$ and $([s]_\simeq, aw) \circlearrowright^* ([r_{i,\vec{y}}]_\simeq, v)$.

- For all $w \in \Sigma^*$ and $i \in \{1, 2, \ldots, n\}$ with $\vec{c}_i \le \psi(w)$, there is a $v \in \Sigma^*$ such that $\vec{c}_i + \psi(v) = \psi(w)$ and $\left(\left[p_{i,\vec{0}}\right]_\simeq, w\right) \circlearrowright^* ([p_{i,\vec{c}_i}]_\simeq, v)$.

- For all $a \in \Sigma$, $w \in \Sigma^*$, $i \in \{1, 2, \ldots, n\}$, and $\vec{y} \in P_i$ with $\psi(a) \leq \vec{y} \leq \psi(aw)$, there is a $v \in \Sigma^*$ such that $\vec{y} + \psi(v) = \psi(aw)$ and $\left(\left[t_{i,\vec{y},\vec{0}} \right]_\approx, aw \right) \circlearrowright^* \left([t_{i,\vec{y},\vec{y}}]_\approx, v \right)$.

Let now $w \in \psi^{-1} \left(\bigcup_{i=1}^n \mathsf{L}(\vec{c}_i, P_i) \right)$. Then, there is an $i \in \{1, 2, \ldots, n\}$ and for each $\vec{y} \in P_i$ a $\lambda_{\vec{y}} \in \mathbb{N}$ such that $\psi(w) = \vec{c}_i + \sum_{\vec{y} \in P_i} \lambda_{\vec{y}} \vec{y}$. Our four statements imply: If $\sum_{\vec{y} \in P_i} \lambda_{\vec{y}} = 0$, we have $([s]_\approx, w) \circlearrowright^* ([q_{i,\vec{c}_i}]_\approx, \lambda)$. If, on the other hand, $\sum_{\vec{y} \in P_i} \lambda_{\vec{y}} > 0$, it holds $([s]_\approx, w) \circlearrowright^* ([p_{i,\vec{c}_i}]_\approx, \lambda)$. Hence, we get $w \in L_R(A)$ in both cases. This shows $\psi^{-1} \left(\bigcup_{i=1}^n \mathsf{L}(\vec{c}_i, P_i) \right) \subseteq L_R(A)$. $\square$

Proposition 8.9 and Example 8.3 show that there is a permutation closed language in **NROWJ** which is not in **MROWJ**. For these two families we get:

Proposition 8.10. *The classes* **MROWJ** *and* **NROWJ** *are incomparable.*

Proof. Let $\Sigma = \{a, b, c\}$ and consider the language

$$L = \left\{ vcaw \mid v, w \in \{b, c\}^{*(\Sigma^*, \Sigma, \lambda_\Sigma)} \land |vcaw|_b = |vcaw|_c \right\}$$
$$\cup \left\{ avc \mid v \in \{b, c\}^{*(\Sigma^*, \Sigma, \lambda_\Sigma)} \land |avc|_b = |avc|_c \right\}$$

and the DFA $A = (\{q_0, q_1, \ldots, q_5\}, \Sigma, \delta, \{q_0\}, \{q_4\})$ with

$$\delta = \{q_0 b \to q_1, q_1 a \to q_2, q_1 c \to q_3, q_3 b \to q_1,$$
$$q_3 a \to q_4, q_4 c \to q_5, q_5 b \to q_4\}.$$

We show $L_R(A) = L$. It is easy to see that for all $w \in L(A)$ we have that $|w|_a = 1$ and $|w|_b = |w|_c > 0$. Let $v, w \in \{b, c\}^{*(\Sigma^*, \Sigma, \lambda_\Sigma)}$. Then, there is an $x \in \{b, c\}^{*(\Sigma^*, \Sigma, \lambda_\Sigma)}$ with $(q_0, vbaw) \circlearrowright_A^{|v|+1} (q_1, awx) \circlearrowright_A (q_2, wx)$ and

$$(q_0, avb) \circlearrowright_A (q_0, vba) \circlearrowright_A^{|v|+1} (q_1, ax) \circlearrowright_A (q_2, x).$$

It follows $L_R(A) \subseteq L$. For all $v, w \in \{b, c\}^{*(\Sigma^*, \Sigma, \lambda_\Sigma)}$ with $|vcaw|_b = |vcaw|_c$ and $|v|_b > 0$ there is an $x \in \{b, c\}^{*(\Sigma^*, \Sigma, \lambda_\Sigma)}$ with $|wx|_b = |wx|_c$ and

$$(q_0, vcaw) \circlearrowright_A^{|v|+1} (q_3, awx) \circlearrowright_A (q_4, wx) \circlearrowright_A^* (q_4, \lambda).$$

For all words $v, w \in \{b, c\}^{*(\Sigma^*, \Sigma, \lambda\Sigma)}$ with $|vcaw|_b = |vcaw|_c$ and $|v|_b = 0$ there exists a word $x \in \{b, c\}^{*(\Sigma^*, \Sigma, \lambda\Sigma)}$ with $|x|_b = |x|_c$ and

$$(q_0, vcaw) \circlearrowright_A^{|v|+2} (q_0, wvca) \circlearrowright_A^{|wv|+1} (q_3, ax) \circlearrowright_A (q_4, x) \circlearrowright_A^* (q_4, \lambda).$$

Let now $v \in \{b, c\}^{*(\Sigma^*, \Sigma, \lambda\Sigma)}$ with $|avc|_b = |avc|_c$. Then, there exists an $x \in \{b, c\}^{*(\Sigma^*, \Sigma, \lambda\Sigma)}$ with $|x|_b = |x|_c$ and

$$(q_0, avc) \circlearrowright_A (q_0, vca) \circlearrowright_A^{|v|+1} (q_3, ax) \circlearrowright_A (q_4, x) \circlearrowright_1^* (q_4, \lambda).$$

This shows $L_R(A) = L$. Hence, L belongs to **ROWJ** and $L \cup \{a\}$ is a member of **MROWJ**.

We show that $L \cup \{a\}$ is not in **NROWJ**. Assume to the contrary that there is an NFA $B = (Q, \Sigma, \gamma, \{s\}, F)$ with $L_R(B) = L \cup \{a\}$. It follows $\gamma(s, a) \neq \emptyset$. For each state $q \in \gamma(s, a)$ set $B_q = (Q, \Sigma, \gamma, \{q\}, F)$ and $L_q = L_R(B_q)$. We have

$$\bigcup_{q \in \gamma(s,a)} L_q = \left\{ vc \mid v \in \{b, c\}^{*(\Sigma^*, \Sigma, \lambda\Sigma)} \wedge |vc|_b = |vc|_c \right\} \cup \{\lambda\}.$$

Thus, there is a $q \in \gamma(s, a)$ with $|L_q \cap \{b^n c^n \mid n \geq 0\}| = \infty$. This implies that there exist $p \in Q$, $f \in F$, $n \geq 0$, and an $m > n$ such that $(q, b^n) \vdash_B^n (p, \lambda)$ and $(p, b^{m-n}c^m) \circlearrowright_B (p, b^{m-n-1}c^m b) \circlearrowright_B^* (f, \lambda)$. We get $(q, b^{m-1}c^m b) \vdash_B^n (p, b^{m-1-n}c^m b) \circlearrowright_B^* (f, \lambda)$, a contradiction. Hence, $L \cup \{a\}$ is not in **NROWJ**. $\qquad\square$

8.3 Nondeterministic ROWJFAs with Spontaneous Transitions

Now, the question arises how the right one-way jumping relation can be generalized to λ-MNFAs. To cope with spontaneous transitions we have the following three different interpretation possibilities for the right one-way jumping relation. Let $A = (Q, \Sigma, \delta, S, F)$ be a λ-MNFA.

Type 1. The *first type of right one-way jumping finite automata* is only allowed to jump over an input symbol if A cannot perform a λ-transition in the current state. So, the binary *right one-way jumping relation of type 1*, symbolically denoted by $\circlearrowright_{A,1}$, on the set of configurations of A is defined as follows. For $p \in Q$, $a \in \Sigma \cup \{\lambda\}$, a state $q \in \delta(p, a)$, and $w \in \Sigma^*$ we have $(p, aw) \circlearrowright_{A,1} (q, w)$. For $p \in Q$, $a \in \Sigma$, and $w \in \Sigma^*$ with $\delta(p, a) = \emptyset = \delta(p, \lambda)$ it holds $(p, aw) \circlearrowright_{A,1} (p, wa)$.

Type 2. If the *second type of right one-way jumping finite automata* has no transition for the current state and the next input symbol, but can perform a λ-transition in the current state, the automaton is allowed to choose if it performs a jump or if it uses a λ-transition. However, if the automaton decides to jump, it has to jump over all the following input symbols that cannot be read in the current state and must read the next input symbol that can be read in the current state. Not until now, the automaton is allowed to perform a λ-transition again. Therefore the *right one-way jumping relation of type 2*, symbolically denoted by $\circlearrowright_{A,2}$, on the set of configurations of A is defined as follows. For $p \in Q$, $q \in \delta(p, \lambda)$, and $w \in \Sigma^*$ it holds $(p, w) \circlearrowright_{A,2} (q, w)$. For $p \in Q$, $a \in \Sigma$, $q \in \delta(p, a)$, $v \in (\Sigma \setminus \Sigma_p)^{*(\Sigma^* \cdot \Sigma \cdot \lambda_\Sigma)}$, and $w \in \Sigma^*$ we get $(p, vaw) \circlearrowright_{A,2} (q, wv)$.

Type 3. If the *third type of right one-way jumping finite automata* has no transition for the current state and the next input symbol, but can perform a λ-transition in the current state, the automaton is allowed to choose if it jumps over the next input symbol or if it uses a λ-transition. Thus, the *right one-way jumping relation of type 3*, symbolically denoted by $\circlearrowright_{A,3}$, on the set of configurations of A is defined as follows. For $p \in Q$, $a \in \Sigma \cup \{\lambda\}$, $q \in \delta(p, a)$, and $w \in \Sigma^*$ we have $(p, aw) \circlearrowright_{A,3} (q, w)$. For $p \in Q$, $a \in \Sigma$ with $\delta(p, a) = \emptyset$, and $w \in \Sigma^*$ holds $(p, aw) \circlearrowright_{A,3} (p, wa)$.

Let $i \in \{1, 2, 3\}$. We also write $\circlearrowright_i$ instead of $\circlearrowright_{A,i}$ if it is clear which automaton we are referring to. The *language accepted by A interpreted as a right one-way jumping finite automaton of type i* is defined to be

$$L_{Ri}(A) = \{ w \in \Sigma^* \mid \exists s \in S, f \in F : (s, w) \circlearrowright_i^* (f, \lambda) \}.$$

This give rise to the language family $\lambda_i\mathbf{MNROWJ}$ ($\lambda_i\mathbf{NROWJ}$, respectively) of all languages accepted by λ-MNFAs (λ-NFAs, respectively) interpreted as right one-way jumping finite automata of type i. For each $i \in \{1,2,3\}$ and each MNFA A it clearly holds $L_{Ri}(A) = L_R(A)$. We also have:

Proposition 8.11. *For all λ-MNFAs A it holds*

$$L(A) \subseteq L_{R1}(A) \subseteq L_{R2}(A) \subseteq L_{R3}(A) \subseteq L_J(A).$$

Proof. The inclusions $L(A) \subseteq L_{R1}(A)$ and $L_{R3}(A) \subseteq L_J(A)$ are immediate from the definitions. It remains to show that $L_{R1}(A) \subseteq L_{R2}(A)$ and $L_{R2}(A) \subseteq L_{R3}(A)$. First, we show *via* induction over n that for all $n \geq 0$, $p \in Q$, $f \in F$, and $w \in \Sigma^*$ the relation $(p,w) \circlearrowright_1^n (f,\lambda)$ implies that there is an $m \leq n$ with $(p,w) \circlearrowright_2^m (f,\lambda)$. The statement is clearly true for $n = 0$. So, let $n > 0$, $p \in Q$, $f \in F$, and $w \in \Sigma^*$ with $(p,w) \circlearrowright_1^n (f,\lambda)$. If there is a $q \in \delta(p,\lambda)$ with $(p,w) \circlearrowright_1 (q,w) \circlearrowright_1^{n-1} (f,\lambda)$, we also have that $(p,w) \circlearrowright_2 (q,w)$ and that there is an $m \leq n-1$ with $(q,w) \circlearrowright_2^m (f,\lambda)$ *via* the induction hypothesis. If there is no $q \in \delta(p,\lambda)$ with $(p,w) \circlearrowright_1 (q,w) \circlearrowright_1^{n-1} (f,\lambda)$, there exist $a \in \Sigma_{\delta,p}$, $r \in \delta(p,a)$, $v \in (\Sigma \setminus \Sigma_{\delta,p})^{*(\Sigma^*,\Sigma,\lambda_\Sigma)}$, and $x \in \Sigma^*$ with $w = vax$ and $(p,vax) \circlearrowright_1^{|v|+1} (r,xv) \circlearrowright_1^{n-|v|-1} (f,\lambda)$. In this case we get $(p,w) = (p,vax) \circlearrowright_2 (r,xv)$ and the existence of an $m \leq n - |v| - 1$ with $(r,xv) \circlearrowright_2^m (f,\lambda)$ because of the induction hypothesis.

Via induction over n, we now prove that for all $n \geq 0$, $p \in Q$, $f \in F$, and $w \in \Sigma^*$ the relation $(p,w) \circlearrowright_2^n (f,\lambda)$ implies $(p,w) \circlearrowright_3^* (f,\lambda)$. The statement is true for $n = 0$. So, let $n > 0$, $p \in Q$, $f \in F$, and $w \in \Sigma^*$ with $(p,w) \circlearrowright_2^n (f,\lambda)$. If there is a $q \in \delta(p,\lambda)$ with $(p,w) \circlearrowright_2 (q,w) \circlearrowright_2^{n-1} (f,\lambda)$, we also have $(p,w) \circlearrowright_3 (q,w)$ and $(q,w) \circlearrowright_3^* (f,\lambda)$ *via* the induction hypothesis. If there is no $q \in \delta(p,\lambda)$ with $(p,w) \circlearrowright_2 (q,w) \circlearrowright_2^{n-1} (f,\lambda)$, there exist $a \in \Sigma_{\delta,p}$, $r \in \delta(p,a)$, $v \in (\Sigma \setminus \Sigma_{\delta,p})^{*(\Sigma^*,\Sigma,\lambda_\Sigma)}$ and $x \in \Sigma^*$ with $w = vax$ and $(p,vax) \circlearrowright_2 (r,xv) \circlearrowright_2^{n-1} (f,\lambda)$. In this case we get $(p,w) = (p,vax) \circlearrowright_3^{|v|+1} (r,xv)$ and $(r,xv) \circlearrowright_3^* (f,\lambda)$ because of the induction hypothesis. $\square$

Example 8.12. Let $\Sigma = \{a, b, c\}$ and consider the λ-NFA

$$A = (\{q_0, q_1, \ldots, q_5\}, \Sigma, \delta, \{q_0\}, \{q_5\})$$

with

$$\delta = \{q_0\lambda \to q_1, q_0a \to q_4, q_1a \to q_2, q_1b \to q_2,$$
$$q_1c \to q_3, q_3a \to q_4, q_4b \to q_5, q_5a \to q_4\}.$$

The automaton A is depicted in Figure 8.1. It is easy to see that that

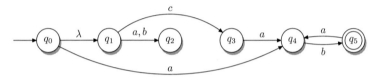

Figure 8.1: The λ-NFA A with $L(A) \subset L_{R1}(A) \subset L_{R2}(A) \subset L_{R3}(A) \subset L_J(A)$.

$$L(A) = \{\lambda, c\} \left(\{ab\}^{*(\Sigma^*, \Sigma, \lambda_\Sigma)} \setminus \{\lambda\} \right).$$

Let

$$L = \{ w \in \{a, b, c\}^* \mid |w|_c = 0 < |w|_a = |w|_b \}.$$

We consider the three interpretations on how to cope with spontaneous transitions in the following. Let $x \in \{a, b\}$, $u, v \in \{b, c\}^{*(\Sigma^*, \Sigma, \lambda_\Sigma)}$, and $w \in \{a, b, c\}^*$.

- In type 1 we have $(q_0, aw) \circlearrowright_1 (q_4, w)$, $(q_0, xw) \circlearrowright_1^2 (q_2, w)$, and also $(q_0, cvaw) \circlearrowright_1^2 (q_3, vaw) \circlearrowright_1^{|v|+1} (q_4, wv)$. This implies

$$L_{R1}(A) = \left\{ w \in \{a\}\{a, b\}^{*(\Sigma^*, \Sigma, \lambda_\Sigma)} \mid |w|_a = |w|_b \right\} \cup \{c\}L.$$

- For type 2 we get $(q_0, vaw) \circlearrowright_2 (q_4, wv)$, $(q_0, xw) \circlearrowright_2^2 (q_2, w)$, and also $(q_0, cvaw) \circlearrowright_2^2 (q_3, vaw) \circlearrowright_2 (q_4, wv)$. That gives us $L_{R2}(A) = \{\lambda, c\}L$.

- In type 3 it holds $(q_0, vaw) \circlearrowright_3^{|v|+1} (q_4, wv)$, $(q_0, vxw) \circlearrowright_3^{|v|+2} (q_2, wv)$, and moreover $(q_0, vcuaw) \circlearrowright_3^{|v|+2} (q_3, uawv) \circlearrowright_3^{|u|+1} (q_4, wvu)$. Hence, the language $L_{R3}(A)$ equals

$$L \cup \Big\{ b^n cw \mid n \geq 0 \wedge w \in \{a,b\}^{*(\Sigma^*, \Sigma, \lambda_\Sigma)} \setminus \{\lambda\} \wedge |b^n cw|_a = |b^n cw|_b \Big\}.$$

Furthermore, we get

$$L_J(A) = \mathsf{perm}(L(A)) = \{ w \in \{a,b,c\}^* \mid |w|_c \leq 1 \leq |w|_a = |w|_b \}.$$

It follows

$$L(A) \subset L_{R1}(A) \subset L_{R2}(A) \subset L_{R3}(A) \subset L_J(A).$$

We study the computational capacity of λ-MNFAs interpreted as right one-way jumping finite automata in the following. Analogously to the proof of Proposition 6.2, we get:

Proposition 8.13. *For all* $i \in \{1,2,3\}$ *it holds* $\lambda_i\mathbf{MNROWJ} \subset \mathbf{NST}(n, n^2)$. $\qquad\square$

If we fix the semantics, λ-NFAs interpreted as right one-way jumping finite automata have the same computational power as λ-MNFAs interpreted as right one-way jumping finite automata:

Proposition 8.14. *For all* $i \in \{1,2,3\}$ *we have that* $\lambda_i\mathbf{NROWJ} = \lambda_i\mathbf{MNROWJ}$.

Proof. Let $A = (Q, \Sigma, \delta, S, F)$ be a λ-MNFA. Assume that $s, d \notin Q$. Then, we define $B = (Q \cup \{s, d\}, \Sigma, \gamma, \{s\}, F)$, where $\gamma|_{Q \times (\Sigma \cup \{\lambda\})} = \delta$, for all $a \in \Sigma$ it holds $\gamma(s, a) = \{d\}$, $\gamma(s, \lambda) = S$, and for all $a \in \Sigma \cup \{\lambda\}$ we have $\gamma(d, a) = \emptyset$. It is easy to see that for all $i \in \{1,2,3\}$ it holds $L_{Ri}(B) = L_{Ri}(A)$. $\qquad\square$

Next, we show that λ-MNFAs interpreted as right one-way jumping finite automata of type 1 are at most as powerful as those of type 2 and those of type 3:

Proposition 8.15. *We have*

$$\lambda_1 \mathbf{MNROWJ} \subseteq \lambda_2 \mathbf{MNROWJ} \cap \lambda_3 \mathbf{MNROWJ}.$$

Proof. Let $A = (Q, \Sigma, \delta, S, F)$ be a λ-MNFA. Assume that $d \notin Q$ and define $B = (Q \cup \{d\}, \Sigma, \gamma, S, F)$, where for all $(q, a) \in (Q \cup \{d\}) \times (\Sigma \cup \{\lambda\})$ it holds

$$\gamma(q, a) = \begin{cases} \emptyset & \text{if } q = d, \\ \delta(q, a) & \text{if } q \in Q \text{ and } \delta(q, \lambda) = \emptyset, \\ \delta(q, a) \cup \{d\} & \text{if } q \in Q \text{ and } \delta(q, \lambda) \neq \emptyset. \end{cases}$$

Then, it holds $L_{R1}(A) = L_{R1}(B) = L_{R3}(B)$. With Proposition 8.11 we also get $L_{R1}(A) = L_{R2}(B)$. $\square$

MNFAs interpreted as right one-way jumping finite automata have the same power as λ-MNFAs interpreted as right one-way jumping finite automata of type 2:

Proposition 8.16. *We have* $\mathbf{MNROWJ} = \lambda_2 \mathbf{MNROWJ}$.

Proof. Let $A = (Q, \Sigma, \delta, S, F)$ be a λ-MNFA. We define the binary relation $\#$ on Q as follows: for all $p, q \in Q$ let $p \# q$ if and only if $q \in \delta(p, \lambda)$. For all $P \subseteq Q$ we set $\ell(P) = \{ q \in Q \mid \exists p \in P : p \#^* q \}$. Define $B = (Q, \Sigma, \gamma, \ell(S), F)$, where for all $(q, a) \in Q \times \Sigma$ it holds $\gamma(q, \lambda) = \emptyset$ and $\gamma(q, a) = \ell(\delta(q, a))$. Then, we have $L_{R2}(B) = L_{R2}(A)$. $\square$

In contrast to the last result, λ-MNFAs interpreted as right one-way jumping finite automata of type 3 are more powerful than MNFAs interpreted as right one-way jumping finite automata:

Proposition 8.17. *It holds* $\mathbf{MNROWJ} \subset \lambda_3 \mathbf{NROWJ}$.

Proof. Consider the λ-NFA

$$A = (\{q_0, q_1, \dots, q_5\}, \{a, b, c\}, \delta, \{q_0\}, \{q_4\})$$

with

$$\delta = \{q_0\lambda \to q_1, q_1a \to q_2, q_2a \to q_3, q_2b \to q_3,$$
$$q_2c \to q_4, q_4a \to q_5, q_4b \to q_5, q_5c \to q_4\}.$$

We show that $L_{R3}(A)$ equals

$$L = \{\, vacw \mid v, w \in \{a, b, c\}^* \wedge |vacw|_a + |vacw|_b = |vacw|_c \,\}$$
$$\cup \{\, cva \mid v \in \{a, b, c\}^* \wedge |cva|_a + |cva|_b = |cva|_c \,\}.$$

Clearly, for all $w \in L(A)$ we have $|w|_a + |w|_b = |w|_c > 0$. Furthermore, for all $w \in L_{R3}(A)$ there is a $v \in \{a, b, c\}^*$ such that acv is a permutation of w and $(q_0, w) \circlearrowleft_3^* (q_1, acv) \circlearrowleft_3^2 (q_4, v)$. It follows $L_{R3}(A) \subseteq L$. Conversely, for all $v, w \in \{a, b, c\}^*$ with $|vacw|_a + |vacw|_b = |vacw|_c$ we have $(q_0, vacw) \circlearrowleft_3^{|v|} (q_0, acwv) \circlearrowleft_3^3 (q_4, wv)$. For all $v \in \{a, b, c\}^*$ with $|cva|_a + |cva|_b = |cva|_c$ it holds $(q_0, cva) \circlearrowleft_3^{|v|+1} (q_0, acv) \circlearrowleft_3^3 (q_4, v)$. Thus, we have $L_{R3}(A) = L$.

Now, we show that L is not in **MNROWJ**. Assume to the contrary that there is an MNFA $B = (Q, \{a, b, c\}, \gamma, S, F)$ with $L_R(B) = L$. For all $s, t \in S$, $p \in Q$, $f \in F$, and $v, w, x \in \{a, b, c\}^*$ with $(s, v) \vdash^{|v|} (p, \lambda)$, $(t, w) \vdash^{|w|} (p, \lambda)$, and $(p, x) \vdash^{|x|} (f, \lambda)$, we clearly have $|v|_a + |v|_b - |v|_c = |w|_a + |w|_b - |w|_c$. For all $p \in Q$ for which there exist states $s \in S$, $f \in F$, and $v, x \in \{a, b, c\}^*$ with $(s, v) \vdash^{|v|} (p, \lambda)$ and $(p, x) \vdash^{|x|} (f, \lambda)$, we set $\Delta(p) = |v|_a + |v|_b - |v|_c$. Let m be the maximum of the set

$$\Big\{\Delta(p) \mid p \in Q \wedge \exists s \in S, f \in F, v, x \in \{a, b, c\}^* :$$
$$((s, v) \vdash^{|v|} (p, \lambda) \wedge (p, x) \vdash^{|x|} (f, \lambda))\Big\}.$$

Because $(ab)^{m+1}aca^{m+2}bc^{3m+5} \in L$, there are $p \in Q$, $s \in S$, $f \in F$, and a $v \in \{a, b, c\}^*$ with $|w|_c = 0$ and

$$(s, (ab)^{m+1}aca^{m+2}bc^{3m+5}) \circlearrowleft_B^{2m+2} (p, aca^{m+2}bc^{3m+5}v) \circlearrowleft_B^* (f, \lambda).$$

We have $\gamma(p, a) = \gamma(p, b) = \emptyset$ and $\gamma(p, c) \neq \emptyset$. So, there exist $q \in \gamma(p, c)$, $r \in Q$, and $n > 0$ with

$$(p, aca^{m+2}bc^{3m+5}v) \circlearrowleft_B^2 (q, a^{m+2}bc^{3m+5}va)$$
$$\circlearrowleft_B^{m+2} (r, bc^{3m+5}vaa^n) \circlearrowleft_B^* (f, \lambda).$$

It follows $\gamma(r, a) = \emptyset$. This gives us

$$(s, (ab)^{m+1}ca^{m+3}bc^{3m+5}) \circlearrowright_B^{2m+2} (p, ca^{m+3}bc^{3m+5}v) \circlearrowright_B (q, a^{m+3}bc^{3m+5}v)$$
$$\circlearrowright_B^{m+2} (r, abc^{3m+5}va^n) \circlearrowright_B (r, bc^{3m+5}va^na) \circlearrowright_B^* (f, \lambda),$$

which implies $(ab)^{m+1}ca^{m+3}bc^{3m+5} \in L$, a contradiction. $\qquad\square$

8.4 Relations to Right-Revolving Finite Automata

We compare right one-way jumping finite automata to right-revolving finite automata as defined by Bensch *et al.* [11]. A *nondeterministic right-revolving finite automaton*, an rr-NFA for short, is a tuple of the form $A = (Q, \Sigma, \delta, \Delta, s, F)$, where Q is the finite *set of states*, Σ is the finite *input alphabet*, δ and Δ are functions from $Q \times \Sigma$ to 2^Q, $s \in Q$ is the *start state*, and $F \subseteq Q$ is the set of *final states*. The automaton A is called a *deterministic right-revolving finite automaton*, an rr-DFA for short, if for all $(p, a) \in Q \times \Sigma$ it holds $|\delta(p, a) \cup \Delta(p, a)| \leq 1$.

A *configuration* of A is an element of $Q \times \Sigma^*$. For $p, q \in Q$, $a \in \Sigma$, and $w \in \Sigma^*$ the automaton A makes a *transition* from configuration (p, aw) to configuration (q, w), denoted by $(p, aw) \vdash_A (q, w)$, if $q \in \delta(p, a)$. If, on the other hand, $q \in \Delta(p, a)$, the automaton A makes a *transition* from configuration (p, aw) to configuration (q, wa), denoted by $(p, aw) \vdash_A (q, wa)$. We also write $\vdash$ instead of $\vdash_A$ if it is clear which automaton we are referring to. The *language accepted by A* is $L(A) = \{ w \in \Sigma^* \mid \exists f \in F : (s, w) \vdash^* (f, \lambda) \}$. The family of all languages accepted by rr-NFAs (rr-DFAs, respectively) is denoted by **NRR** (**DRR**, respectively).

By definition, we have $\mathbf{ROWJ} \subseteq \mathbf{DRR}$. A sketch of a proof that it holds $\mathbf{ROWJ} \subset \mathbf{DRR}$ was given by Chigahara *et al.* [14]. Originally, right-revolving finite automata were defined with λ-transitions, but it was shown by Bensch *et al.* that the resulting language families are the same if you are not allowed to use λ-transitions [11]. Hence, we get that it holds λ_3**MNROWJ** $\subseteq$ **NRR**. Bensch *et al.* have shown that the permutation closed deterministic context-free language $\{\, w \in \{a,b\}^* \mid |w|_a = |w|_b \vee |w|_b = 0 \,\}$, which clearly is in **MROWJ**, is *not* in **DRR** [11]. All languages in λ_3**MNROWJ** are semilinear. Also, all the examples of languages from **NRR** given by Bensch *et al.* are semilinear [11]. However, there is a non-semilinear language even in **pDRR**:

Proposition 8.18. *We have*

$$\{\, w \in \{a,b\}^* \mid |w|_b = 1 \wedge \exists n \geq 0 : |w|_a = 2^n \,\} \in \mathbf{DRR}.$$

Proof. Consider the rr-DFA $A = (\{q_0, q_1, \ldots, q_5\}, \{a, b\}, \delta, \Delta, q_0, \{q_3\})$, where

$$\delta(q_1, a) = \{q_2\},\ \delta(q_2, b) = \{q_3\},\ \delta(q_4, a) = \{q_5\},$$
$$\Delta(q_0, a) = \{q_0\},\ \Delta(q_0, b) = \{q_1\},\ \Delta(q_2, a) = \{q_4\},$$
$$\Delta(q_4, b) = \{q_1\},\ \Delta(q_5, a) = \{q_4\},$$

and all other values of δ and Δ are $\emptyset$. For all $w \in L(A)$ we clearly have $|w|_b = 1$ and $|w|_a \geq 1$.

We show *via* induction over m that for all $m \geq 0$ and $w \in \{a,b\}^*$ with $|w|_b = 1$ and $|w|_a \leq 2^m$ it holds $w \in L(A)$ if and only if there is an $n \geq 0$ with $|w|_a = 2^n$. It is easy to see that $ab, ba \in L(A)$. Let now $m \geq 1$ and $w \in \{a,b\}^*$ with $|w|_b = 1$ and $2^{m-1} < |w|_a \leq 2^m$. If $|w|_a$ is odd, it holds $(q_0, w) \vdash^* \left(q_5, ba^{(|w|_a-1)/2}\right)$, which implies $w \notin L(A)$. If $|w|_a$ is even and $|w|_a < 2^m$, we have $(q_0, w) \vdash^* \left(q_1, a^{|w|_a/2}b\right)$, with $2^{m-2} < |w|_a/2 < 2^{m-1}$. The induction hypothesis gives us $w \notin L(A)$. Finally, if $|w|_a = 2^m$, it holds $(q_0, w) \vdash^* \left(q_1, a^{2^{m-1}}b\right)$. In this case the induction hypothesis tells us $w \in L(A)$. That shows

$$L(A) = \{\, w \in \{a,b\}^* \mid |w|_b = 1 \wedge \exists n \geq 0 : |w|_a = 2^n \,\}.$$

$\square$

8.5 Relations to Finite-State Acceptors With Translucent Letters

Right one-way jumping finite automata are compared to finite-state acceptors with translucent letters as defined by Nagy and Otto [58]. A *finite-state acceptor with translucent letters*, an NFAwtl for short, is a tuple $A = (Q, \Sigma, \tau, S, F, \delta)$, where Q is the finite *set of states*, Σ is the finite *input alphabet*, $\tau : Q \to 2^{\Sigma}$ is the *translucency mapping*, $S \subseteq Q$ is the set of *start states*, $F \subseteq Q$ is the set of *final states*, and $\delta : Q \times \Sigma \to 2^Q$ is the *transition function*. For each state $q \in Q$ the letters from $\tau(q)$ are *translucent* in state q. The automaton A is called *deterministic*, a DFAwtl for short, if $|S| = 1$ and for all $(q, a) \in Q \times \Sigma$ it holds $|\delta(q, a)| \leq 1$.

A *configuration* of A is an element of $Q \times \Sigma^*$. For $p \in Q$, $v \in (\tau(p))^{*(\Sigma^*, \Sigma, \lambda_\Sigma)}$, $a \in \Sigma \setminus \tau(p)$, $q \in \delta(p, a)$, and $w \in \Sigma^*$ the automaton A makes a *transition* from configuration (p, vaw) to configuration (q, vw), denoted by $(p, vaw) \vdash_A (q, vw)$ or just $(p, vaw) \vdash (q, vw)$ if it is clear which automaton we are referring to. The *language accepted by A* is

$$L(A) = \left\{ w \in \Sigma^* \mid \exists s \in S, f \in F, v \in (\tau(f))^{*(\Sigma^*, \Sigma, \lambda_\Sigma)} : (s, w) \vdash^* (f, v) \right\}.$$

The automaton A is in *normal form* if for all words $w \in \Sigma^*$, states $s \in S$ and $f \in F$, and words $v \in (\tau(f))^{*(\Sigma^*, \Sigma, \lambda_\Sigma)}$ with $(s, w) \vdash_B^* (f, v)$ it holds $v = \lambda$. The family of all languages accepted by an NFAwtl (DFAwtl, respectively) is denoted by **NTRANS** (**DTRANS**, respectively). The next proposition can be easily seen.

Proposition 8.19. *For every NFAwtl A there is an NFAwtl $B = (Q, \Sigma, \tau, S, F, \delta)$ with $L(B) = L(A)$ such that for all $(q, a) \in Q \times \Sigma$ we have $\delta(q, a) = \emptyset$ if and only if $a \in \tau(q)$. If A is a DFAwtl, so is B. If A is in normal form, B is also.* $\square$

It follows from the definitions that each language from **NTRANS** contains a Parikh-equivalent regular sublanguage. Nagy and Otto showed that each permutation closed semilinear language is in **NTRANS** while the permutation closed context-free language $\{ w \in \{a, b\}^* \mid |w|_b \in \{|w|_a, 2 \cdot |w|_a\} \}$, which clearly is in **MROWJ**, is not in **DTRANS** [58]. On the other hand, it is not hard to see that for all $n > 0$ the permutation closed language

$$\{\, w \in \{a_1, a_2, \ldots, a_n\}^* \mid |w|_{a_1} = |w|_{a_2} = \cdots = |w|_{a_n} \,\}$$

is in **DTRANS**. The family **DTRANS** is clearly closed under the operation of complementation. Hence, the permutation closed language $\{\, w \in \{a, b\}^* \mid |w|_a \neq |w|_b \,\}$, which is not in **MROWJ** by Example 8.3, is in **DTRANS**. Nagy and Otto showed that every NFAwtl is equivalent to an NFAwtl in normal form [58]. It was stated as an open problem by them if every DFAwtl is equivalent to a DFAwtl in normal form. Every permutation closed language that is accepted by a DFAwtl in normal form is obviously in **ROWJ**. Theorem 6.4 answers the open problem by Nagy and Otto:

Corollary 8.20. *The language* $\{\, w \in \{a, b\}^* \mid |w|_a \neq |w|_b \,\}$, *which is in* **DTRANS**, *is not accepted by a DFAwtl in normal form.* □

For each DFA $A = (Q, \Sigma, \delta, \{s\}, F)$ such that $L_R(A) = \mathsf{perm}(L_R(A))$ and for all pairs $(f, a) \in F \times \Sigma$ the value $\delta(f, a)$ is defined, it clearly holds $L_R(A) \in$ **DTRANS**. So, the next result gives us $\mathbf{pROWJ} \subset \mathbf{pDTRANS}$.

Proposition 8.21. *For each* $L \in \mathbf{pROWJ}$ *there is a DFA* $A = (Q, \Sigma, \delta, \{s\}, F)$ *such that* $L_R(A) = L$ *and for all* $(f, a) \in F \times \Sigma$ *the value* $\delta(f, a)$ *is defined.*

Proof. Let $A = (Q, \Sigma, \delta, \{s\}, F)$ be a DFA with $L_R(A) = \mathsf{perm}(L_R(A))$. Consider the DFA

$$B = (Q \cup \{\, q_a \mid q \in Q \land a \in \Sigma \,\}, \Sigma, \gamma, \{s\}, F) \,,$$

where γ is defined as follows. For all $q \in Q$ and $a \in \Sigma$ such that $\delta(q, a)$ is defined, it holds that $\gamma(q, a) = \delta(q, a)$. On the other hand, for all states $q \in Q$ and symbols $a \in \Sigma$ such that the value $\delta(q, a)$ is undefined, we have $\gamma(q, a) = q_a$. For all $q \in Q$ and $a, b \in \Sigma$ such that $\delta(q, b)$ is undefined, the value $\gamma(q_a, b)$ is undefined. Moreover, for all states $q \in Q$

and $a, b \in \Sigma$ such that $\delta(q, b)$ is defined and $\delta(\delta(q, b), a)$ is undefined, it holds $\gamma(q_a, b) = (\delta(q, b))_a$. For all $q \in Q$ and $a, b \in \Sigma$ such that $\delta(q, b)$ is defined and $\delta(\delta(q, b), a)$ is also defined, we have $\gamma(q_a, b) = \delta(\delta(q, b), a)$. It is easy to see that $L_R(B) = L_R(A)$. □

The families **DTRANS** and **NTRANS** are both incomparable to all our language families induced by right one-way jumping finite automata, as the next two propositions show.

Proposition 8.22. *The class* **DTRANS**$\backslash\lambda_3$**MNROWJ** *is not empty.*

Proof. Consider the DFAwtl

$$A = (\{q_0, q_1, q_2, q_3\}, \{a, b\}, \tau, \{q_0\}, \{q_3\}, \delta),$$

where $\tau(q_0) = \emptyset$, $\tau(q_1) = \{a\}$, $\tau(q_2) = \{b\}$, $\tau(q_3) = \emptyset$, $\delta(q_0, a) = \delta(q_3, a) = \{q_1\}$, $\delta(q_0, b) = \delta(q_3, b) = \{q_2\}$, $\delta(q_1, b) = \{q_0\}$, $\delta(q_2, a) = \{q_3\}$, and $\delta(q_1, a) = \delta(q_2, b) = \emptyset$. The automaton A is depicted in Figure 8.2. It is obvious that for all $w \in \{a, b\}^*$ we have $|w|_a = |w|_b$ if and only

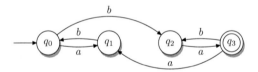

Figure 8.2: The DFAwtl A accepting the language $\{ w \in \{a, b\}^*\{a\} \mid |w|_a = |w|_b \}$.

if $(q_0, w) \vdash^* (q_0, \lambda)$ or $(q_0, w) \vdash^* (q_3, \lambda)$. Let $w \in \{a, b\}^*$ with $|w|_a = |w|_b$. There are $p_1, p_2, \ldots, p_{|w|/2-1} \in \{q_0, q_3\}$ and $w_1, w_2, \ldots, w_{|w|/2-1} \in \{a, b\}^*$ such that

$$(q_0, w) \vdash^2 (p_1, w_1) \vdash^2 (p_2, w_2) \vdash^2 \cdots \vdash^2 (p_{|w|/2-1}, w_{|w|/2-1})$$

and for all $i \in \{1, 2, \ldots, |w|/2-1\}$ it holds $|w_i|_a = |w_{i-1}|_a - 1$ and $|w_i|_b = |w_{i-1}|_b - 1$, where $w_0 = w$. If $w \in \{a, b\}^*\{a\}$, we get $w_{|w|/2-1} = ba$. If, on the other hand, $w \in \{a, b\}^*b$, it holds $w_{|w|/2-1} = ab$. This proves $L(A) =$

$\{\, w \in \{a,b\}^*\{a\} \mid |w|_a = |w|_b \,\}$. We show $L(A) \notin \lambda_3\mathbf{NROWJ}$. Assume to the contrary that there is a λ-NFA $B = (Q, \{a,b\}, \gamma, \{s\}, F)$ with $L_{R3}(B) = L(A)$. Since $\{\, b^n a^n \mid n > 0 \,\} \subset L(A)$, there are $m, n \in \mathbb{N}$ with $m > n$, $p \in Q$, and $f \in F$ with $(s, b^n) \vdash_B^* (p, \lambda)$ and $(p, b^{m-n}a^m) \circlearrowleft_{B,3}$ $(p, b^{m-n-1}a^m b) \circlearrowleft_{B,3}^* (f, \lambda)$. From $(s, b^{m-1}a^m b) \vdash_B^* (p, b^{m-1-n}a^m b) \circlearrowleft_{B,3}^*$ (f, λ), we get $b^{m-1}a^m b \in L_{R3}(B)$, a contradiction. □

Proposition 8.23. *The class* $\mathbf{ROWJ} \setminus \mathbf{NTRANS}$ *is not empty.*

Proof. Let $\Sigma = \{a, b, c\}$ and consider the DFA

$$A = (\{q_0, q_1, q_2, q_3\}, \Sigma, \delta, \{q_0\}, \{q_2\}),$$

where

$$\delta = \{q_0 c \to q_1, q_1 a \to q_1, q_1 b \to q_1, q_1 c \to q_2, q_2 a \to q_3, q_3 b \to q_2\}.$$

The DFA A is depicted in Figure 8.3. It should be clear from the construc-

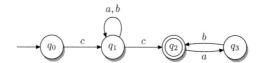

Figure 8.3: The DFA A with $L_R(A) \notin \mathbf{NTRANS}$.

tion that

$$L_R(A) = \left\{\, ucvcw \;\middle|\; u, v, w \in \{a, b\}^{*(\Sigma^*, \Sigma, \lambda_\Sigma)} \wedge |uw|_a = |uw|_b \,\right\}.$$

It remains to show $L_R(A) \notin \mathbf{NTRANS}$. Assume to the contrary that the NFAwtl $B = (Q, \Sigma, \tau, S, F, \gamma)$ accepts $L_R(A)$. Because $\{\, a^n ccb^n \mid n \geq 0 \,\} \subset L_R(A)$, there are $m, n \in \mathbb{N}$ with $m > n$, $s \in S$, $p \in Q$, $f \in F$, and $w \in (\tau(f))^{*(\Sigma^*, \Sigma, \lambda_\Sigma)}$ with $(s, a^n) \vdash_B^n (p, \lambda)$, $a \in \tau(p)$, and $(p, a^{m-n}ccb^m) \vdash_B^*$ (f, w). Clearly, $\tau(p) \subset \{a, b, c\}$. Assume $c \notin \tau(p)$. Then, there is a $q \in$ $\gamma(p, c)$ with $(q, a^{m-n}cb^m) \vdash_B^* (f, w)$. It follows

$$\left(s, a^{m-1}cacb^m\right) \vdash_B^n \left(p, a^{m-1-n}cacb^m\right) \vdash_B \left(q, a^{m-n}cb^m\right) \vdash_B^* (f, w),$$

which is false. Hence, we have $\tau(p) = \{a, c\}$. So, there is an $r \in \delta(p, b)$ fulfilling $(r, a^{m-n}ccb^{m-1}) \vdash_B^* (f, w)$. This gives us

$$\left(s, a^m cbcb^{m-1}\right) \vdash_B^n \left(p, a^{m-n}cbcb^{m-1}\right) \vdash_B \left(r, a^{m-n}ccb^{m-1}\right) \vdash_B^* (f, w),$$

a contradiction. □

To get complexity results about languages accepted by finite-state acceptors with translucent letters, we use the following type of counters. Let

$$\mathscr{C} = \{\, f : \mathbb{N} \to \{-1, 0, 1\} \mid \exists n \in \mathbb{N} : \forall m \in \mathbb{N} : (m \geq n \Rightarrow f(m) = 0)\,\}$$

and $0_\mathscr{C} \in \mathscr{C}$ be the constant 0-function. For an $f \in \mathscr{C}$ let $\mathsf{val}(f) = \Sigma_{n \geq 0} f(n) 2^n$. For every $f \in \mathscr{C}$ we have $\mathsf{val}(f) = 0$ if and only if $f = 0_\mathscr{C}$. For every $f \in \mathscr{C} \setminus \{0_\mathscr{C}\}$ it holds $\mathsf{val}(f) > 0$ if and only if

$$f\left(\max\left(\{\, n \in \mathbb{N} \mid f(n) \neq 0\,\}\right)\right) = 1.$$

Define the maps $\mathsf{inc}, \mathsf{dec} : \mathscr{C} \to \mathscr{C}$ as follows. For all $f \in \mathscr{C}$ and $n \in \mathbb{N}$ the value $\mathsf{inc}(f)(n)$ is

$$\begin{cases} 0 & \text{if for all } m \in \mathbb{N} \text{ with } m \leq n \text{ it holds } f(m) = 1, \\ f(n) + 1 & \text{if } f(n) \leq 0 \text{ and for all } m \in \mathbb{N} \text{ it holds } m \geq n \text{ or } f(m) = 1, \\ f(n) & \text{otherwise,} \end{cases}$$

and $\mathsf{dec}(f)(n)$ is

$$\begin{cases} 0 & \text{if for all } m \in \mathbb{N} \text{ with } m \leq n \text{ it holds } f(m) = -1, \\ f(n) - 1 & \text{if } f(n) \geq 0 \text{ and for all } m \in \mathbb{N} \text{ it holds } m \geq n \text{ or } f(m) = -1, \\ f(n) & \text{otherwise.} \end{cases}$$

For all $f \in \mathscr{C}$ we get $\mathsf{val}(\mathsf{dec}(f)) = \mathsf{val}(f) - 1$ and $\mathsf{val}(\mathsf{inc}(f)) = \mathsf{val}(f) + 1$. Let $n \geq 0$ and $f_0, f_1, \ldots, f_n \in \mathscr{C}$ with $f_0 = 0_\mathscr{C}$ such that for all $i \in \{1, 2, \ldots, n\}$ it holds $f_i = \mathsf{inc}\,(f_{i-1})$ or $f_i = \mathsf{dec}\,(f_{i-1})$. Then, we have

$$\sum_{i=1}^{n} |\{\, m \in \mathbb{N} \mid f_i(m) \neq f_{i-1}(m)\,\}|$$
$$= \sum_{m \geq 0} |\{\, i \in \{1, 2, \ldots, n\} \mid f_i(m) \neq f_{i-1}(m)\,\}| \leq \sum_{m \geq 0} n/2^m = 2n.$$

We use these counters now to get complexity results. It was stated by Nagy and Otto that **NTRANS** $\subseteq$ **NST**(n, n^2) [58]. Kovács and Nagy showed that **NTRANS** $\subseteq$ **NTIME**$(n \log(n))$ and also **DTRANS** $\subseteq$ **DTIME**$(n \log(n))$ [57]. We can improve this:

Proposition 8.24. *Let* $f : \mathbb{N} \to \mathbb{N}$ *fulfil that for all* $n > 0$ *it holds that* $\log(n) \leq f(n) \leq n$. *Assume furthermore that there is a deterministic Turing machine with a read-only input tape, multiple working tapes, and an output tape, such that for every input* $n \in \mathbb{N}$, *encoded in binary,* $O(f(n))$ *working and output tape cells are visited during the computation, the machine halts after* $O(n^2/f(n))$ *steps, and the output is* $f(n)$, *encoded in binary. Then, it holds* **NTRANS** $\subset$ **NST**$(f(n), n^2/f(n))$ *and* **DTRANS** $\subset$ **DST**$(f(n), n^2/f(n))$.

Proof. The non-semilinear language $\{\, a^{m^2} \mid m \geq 0 \,\}$ is in **DST**$(\log(n), n)$. We show the inclusion **DTRANS** $\subseteq$ **DST**$(f(n), n^2/f(n))$. The nondeterministic case is shown analogously. So, let $A = (Q, \Sigma, \tau, \{s\}, F, \delta)$ be a DFAwtl such that for all $(q, a) \in Q \times \Sigma$ we have $\delta(q, a) = \emptyset$ if and only if $a \in \tau(q)$. We construct a deterministic multi-tape Turing machine T which has a read-only input tape such that $L(T) = L(A)$ and for every input $w \in \Sigma^*$ the heads of T visit $O(f(|w|))$ working tape cells during the computation and T halts after $O(|w|^2/f(|w|))$ steps. Let $w \in \Sigma^*$ be an input of T. First of all, T computes a binary encoding of $f(|w|)$ and saves it on a working tape. For every $a \in \Sigma$ there is a counter c_a, whose value at the beginning is the position of the first occurrence of a in w. Furthermore, for every $(a, b) \in \Sigma^2$ there is a counter $c_{a,b}$, whose value is always the difference between the value of c_a and the value of c_b. For every $a \in \Sigma$ there is a tape t_a consisting of $f(|w|)$ cells, which at the beginning contains the factor of length $f(|w|)$ of w that begins with the first occurrence of a in w. For every $a \in \Sigma$ the head of t_a starts at the first symbol. Assume that during the computation, A is in state $q \in Q$. Then, T determines the $a \in \Sigma \setminus \tau(q)$ such that for all $b \in \Sigma \setminus \tau(q)$ the value of $c_{b,a}$ is positive. The current state of A is set to the only element of $\delta(q, a)$. Then, the head of t_a is moved to the next occurrence of a. For every cell that the head of t_a moves to the right, the value of c_a is incremented by 1. After that, the next step of the computation of A is simulated. Whenever, for an $a \in \Sigma$, the head of t_a moves to the end of t_a, the first occurrence of a in w whose position is bigger than the value of c_a is detected, the value of c_a is set to the position of this occurrence, the factor of length $f(|w|)$ of w that begins with this occurrence of a is copied to t_a, and the head of t_a is set to the beginning of the tape. This way, T simulates the computation of A on the input w in space $O(f(|w|))$. For each $a \in \Sigma$ the content of t_a is updated at most $|w|/f(|w|)$ times. For every update the correct factor of w has to be detected. This can be done in $O(|w|)$ steps. □

From Proposition 8.24 it directly follows:

Proposition 8.25. *It holds* $\mathbf{NTRANS} \subset \mathbf{NST}(\log(n), n^2/\log(n))$ *and also* $\mathbf{DTRANS} \subset \mathbf{DST}(\log(n), n^2/\log(n))$. *For all* $p \in \mathbb{Q}$ *with* $0 < p \leq 1$ *we have* $\mathbf{NTRANS} \subset \mathbf{NST}(n^p, n^{2-p})$ *and also* $\mathbf{DTRANS} \subset \mathbf{DST}(n^p, n^{2-p})$. □

Next, we want to show a matching lower bound for Proposition 8.24. In the book *Komplexitätstheorie* by Wolfgang J. Paul it is proven that for the alphabet $\Sigma = \{a, b, c\}$ and all functions $f, g : \mathbb{N} \to \mathbb{N}$ with $f(n) \cdot g(n) \in o(n^2)$ it holds

$$\left\{ wc^{|w|}w \ \middle|\ w \in \{a,b\}^{*(\Sigma^*, \Sigma, \lambda_\Sigma)} \right\} \notin \mathbf{DST}(f(n), g(n)) \text{ [63, Satz 4.38]}.$$

Adapting the mentioned proof, we get:

Proposition 8.26. *Let* Σ *and* Γ *be alphabets with* $|\Sigma| \geq 2$ *and* $\Sigma \subseteq \Gamma$, $m > 0$, *and the map* $\varphi : \Sigma^* \to \Gamma^*$ *be injective such that for all* $w \in \Sigma^*$ *we have* $|\varphi(w)| = m \cdot |w|$. *Then, for all* $f, g : \mathbb{N} \to \mathbb{N}$ *with* $f(n) \cdot g(n) \in o(n^2)$ *it holds*

$$\left\{ (\iota_\Gamma \circ \iota_{\Sigma \to \Gamma})^*(w) \cdot \varphi(w) \mid w \in \Sigma^* \right\} \notin \mathbf{NST}(f(n), g(n)).$$

Proof. Let $k \geq 1$ and $M = (Q, \Gamma, \Delta, \delta, s_0, B, F)$ be a nondeterministic Turing machine with a read-only input tape and k working tapes such that for every computation the machine starts and halts in a situation where the head of the input tape stands on the cell that is the left neighbour of the first input symbol. A *generalized state* of M is an element of $Q \times (\Delta^* \setminus \{\lambda\})^k \times (\mathbb{N}_{>0})^k$. When the machine M is in generalized state $(q, (w_1, w_2, \ldots, w_k), (n_1, n_2, \ldots, n_k))$ this means that M is in state q and for all numbers $i \in \{1, 2, \ldots, k\}$ the content of the i-th working tape is w_i and the head of this tape stands on the n_i-th symbol of w_i. For all computations C of M and all $n \in \mathbb{N}$ the *crossing sequence* $\mathrm{CS}(C, n)$ is the sequence of generalized states that M is in whenever during the computation C the head of the input tape has just crossed the border between the n-th symbol of the input and the $(n+1)$-th symbol. Let $|\mathrm{CS}(C, n)|$ be the number of generalized states in this crossing sequence. The following claim is not hard to see.

Claim. For all $u, v, w, x \in \Gamma^*$ and all computations C on input uv and D on input wx such that $\mathrm{CS}(C, |u|) = \mathrm{CS}(D, |w|)$, there is a canonical computation on input ux that is accepting if and only if C is accepting. □

Let $f : \mathbb{N} \to \mathbb{N}$ and $g : \Gamma^* \to \mathbb{N}$ be functions such that for every computation on input $w \in \Gamma^*$ the machine M visits at most $f(|w|)$ working tape cells and halts after at most $g(w)$ steps. For every $n \in \mathbb{N}$ the number of generalized states that M can reach during computations that start on an input of length n is smaller than

$$G(n) = |Q| \cdot (|\Delta| + 1)^{k \cdot f(n)} \cdot f(n)^k.$$

Define $\iota : \Sigma^* \to \Gamma^*$ as $\iota = (\iota_\Gamma \circ \iota_{\Sigma \to \Gamma})^*$ and assume $L(M) = \{ \iota(w) \cdot \varphi(w) \mid w \in \Sigma^* \}$ now. For every $w \in \Sigma^*$ let $C(w)$ be an accepting computation of M on the input $\iota(w) \cdot \varphi(w)$. Because of the claim, for every $w \in \Sigma^*$ and every $i \in \{0, 1, \ldots, |w|\}$ we have

$$\left| \{ v \in \Sigma^* \mid |v| = |w| \wedge \mathrm{CS}(C(v), i) = \mathrm{CS}(C(w), i) \} \right| \leq |\Sigma|^{|w|-i}.$$

For all $n, i \in \mathbb{N}$ set

$$p(i, n) = |\Sigma|^{-n} \cdot \sum_{\substack{w \in \Sigma^*, \\ |w|=n}} |\mathrm{CS}(C(w), i)|.$$

For all $n \in \mathbb{N}$ and $i \in \{0, 1, \ldots, n\}$ we have

$$(G((m+1)n) + 1)^{\frac{|\Sigma|}{|\Sigma|-1} \cdot p(i,n)}$$

$$> \left| \left\{ \mathrm{CS}(C(w), i) \,\middle|\, w \in \Sigma^* \wedge |w| = n \wedge |\mathrm{CS}(C(w), i)| \leq \frac{|\Sigma|}{|\Sigma| - 1} \cdot p(i,n) \right\} \right|$$

$$\geq |\Sigma|^{i-n} \cdot \left| \left\{ w \in \Sigma^* \,\middle|\, |w| = n \wedge |\mathrm{CS}(C(w), i)| \leq \frac{|\Sigma|}{|\Sigma| - 1} \cdot p(i,n) \right\} \right|$$

$$> |\Sigma|^{i-1},$$

which gives us $p(i, n) > \frac{(|\Sigma|-1)(i-1)}{|\Sigma| \cdot \log_{|\Sigma|}(G((m+1)n)+1)}$. For all $n \in \mathbb{N}$ it holds

$$\max \left(\left\{ g\left(\iota(w) \cdot \varphi(w) \right) \mid w \in \Sigma^* \wedge |w| = n \right\} \right)$$

$$\geq |\Sigma|^{-n} \cdot \sum_{\substack{w \in \Sigma^*, \\ |w|=n}} g\left(\iota(w) \cdot \varphi(w) \right)$$

$$\geq |\Sigma|^{-n} \cdot \sum_{\substack{w \in \Sigma^*, \\ |w|=n}} \sum_{i=0}^{n} |\mathrm{CS}(C(w), i)|$$

$$= \sum_{i=0}^{n} p(i, n)$$

$$> \sum_{i=1}^{n-1} \frac{(|\Sigma| - 1)\, i}{|\Sigma| \cdot \log_{|\Sigma|}(G((m+1)n) + 1)}$$

$$\geq \frac{(|\Sigma| - 1)\, n(n - 1)}{2 \cdot |\Sigma| \cdot (\log_{|\Sigma|}(G((m+1)n)) + 1)},$$

where the last term equals

$$(|\Sigma| - 1)\, n(n - 1) \Big(2 \cdot |\Sigma| \cdot \Big(k \cdot \log_{|\Sigma|}(|\Delta| + 1) \cdot f((m+1)n)$$

$$+ k \cdot \log_{|\Sigma|}(f((m+1)n)) + \log_{|\Sigma|}(|Q|) + 1 \Big) \Big)^{-1},$$

which is in $\Theta\left(n^2 / f((m+1)n)\right)$. □

The following two propositions are consequences of Proposition 8.26.

Proposition 8.27. *For all $f, g : \mathbb{N} \to \mathbb{N}$ with $f(n) \cdot g(n) \in o(n^2)$ and all alphabets Σ with $|\Sigma| \geq 2$ the languages $\{ ww^R \mid w \in \Sigma^* \}$ and $\{ w^2 \mid w \in \Sigma^* \}$ are both not in $\mathbf{NST}(f(n), g(n))$.* □

We have a matching lower bound for Proposition 8.24 over an alphabet of cardinality four:

Proposition 8.28. *There are $L_1, L_2 \in 2^{\{a,b,c,d\}^*}$ with $L_1 \in$ **ROWJ** and $L_2 \in$ **DTRANS** such that for all $i \in \{1,2\}$ and all functions $f, g :$ $\mathbb{N} \to \mathbb{N}$ with $f(n) \cdot g(n) \in o(n^2)$ it holds $L_i \notin$ **NST**$(f(n), g(n))$.*

Proof. Let $\Sigma = \{a, b, c, d\}$, $Q = \{q_0, q_1, q_2, q_3\}$, and A_1 be the DFA given by $(Q, \Sigma, \delta, \{q_0\}, \{q_0\})$, where δ is defined as follows. We have $\delta(q_0, a) = \{q_1\}$, $\delta(q_0, b) = \{q_2\}$, $\delta(q_1, c) = \{q_0\}$, $\delta(q_1, d) = \{q_3\}$, $\delta(q_2, c) = \{q_3\}$, and $\delta(q_2, d) = \{q_0\}$. For all other arguments the value is the empty set. It is not hard to see that

$$L := L_R(A_1) \cap \{a, b\}^{*(\Sigma^*, \cdot_\Sigma, \lambda_\Sigma)} \cdot \{c, d\}^{*(\Sigma^*, \cdot_\Sigma, \lambda_\Sigma)}$$
$$= \left\{ w \cdot h(w) \mid w \in \{a, b\}^{*(\Sigma^*, \cdot_\Sigma, \lambda_\Sigma)} \right\},$$

where $h : \Sigma^* \to \Sigma^*$ is a monoid homomorphism fulfilling $h(a) = c$ and $h(b) = d$. Now, consider the DFAwtl $A_2 = (Q, \Sigma, \tau, \{q_0\}, \{q_0\}, \delta)$, where for all $(q, a) \in (Q \setminus \{q_0\}) \times \Sigma$ we have $a \in \tau(q)$ if and only if $\delta(q, a) = \emptyset$. Furthermore it holds $\tau(q_0) = \emptyset$. We get

$$L(A_2) \cap \{a, b\}^{*(\Sigma^*, \cdot_\Sigma, \lambda_\Sigma)} \cdot \{c, d\}^{*(\Sigma^*, \cdot_\Sigma, \lambda_\Sigma)} = L.$$

Let $f, g : \mathbb{N} \to \mathbb{N}$ be functions with $f(n) \cdot g(n) \in o(n^2)$. If we had

$$\{L_R(A_1), L(A_2)\} \cap \mathbf{NST}(f(n), g(n)) \neq \emptyset,$$

we would also have $L \in \mathbf{NST}(f(n), g(n))$, which is false by Proposition 8.26. □

For ternary alphabets we get:

Proposition 8.29. *Every subset of $\{a, b, c\}^*$ that is in **NTRANS** is in **NST**$(\log(n), n)$. Every subset of $\{a, b, c\}^*$ that is in **DTRANS** is in **DST**$(\log(n), n)$.*

Proof. We show that every subset of $\{a, b, c\}^*$ that is in **DTRANS** is in **DST**$(\log(n), n)$. Again, the nondeterministic case is shown analogously.

Hence, consider a DFAwtl $A = (Q, \{a, b, c\}, \tau, \{s\}, F, \delta)$ such that for all $(q, d) \in Q \times \{a, b, c\}$ we have $\delta(q, d) = \emptyset$ if and only if $d \in \tau(q)$. We construct a deterministic multi-tape Turing machine T which has a read-only input tape such that $L(T) = L(A)$ and for every input $w \in \{a, b, c\}^*$ the heads of T visit $O(\log(|w|))$ working tape cells during the computation and T halts after $O(|w|)$ steps. Let $n \geq 0$ and $d_1, d_2, \ldots, d_n \in \{a, b, c\}$ and consider the input $d_1 d_2 \cdots d_n$ of T. For every $d \in \{a, b, c\}$ there is a counter C_d, whose initial value is $|d_1 d_2 \cdots d_n|_d$. If $n > 0$, set $e_1 = d_1$. If $n = 0$, set $e_1 = a$. Set

$$m = \max \left(\{\, i \in \{1, 2, \ldots, n\} \mid \forall j \in \{1, 2, \ldots, i\} : d_j = e_1 \,\} \right).$$

If $m < n$, set $e_2 = d_{m+1}$. If $m = n$ and $e_1 \neq a$, set $e_2 = a$. If $m = n$ and $e_1 = a$, set $e_2 = b$. Let e_3 be the only element of $\{a, b, c\} \setminus \{e_1, e_2\}$. There are two counters C_1 and C_2. The initial value of C_1 is m. The initial value of C_2 is 0. At the beginning, set the head of the input tape of T to the $(m + 1)$-th symbol of the input. Assume that during the computation, A is in state $q \in Q$. If $\tau(q) = \{a, b, c\}$, the machine T halts. Assume $\tau(q) \subset \{a, b, c\}$ now and set $i = \min(\{\, j \in \{1, 2, 3\} \mid e_j \notin \tau(q) \,\})$. If the value of C_{e_i} is 0, the machine T halts. Assume that the value of C_{e_i} is positive now. In this situation, the value of C_{e_i} is decreased by 1. The current state of A is set to the only element of $\delta(q, e_i)$. If $i = 3$, the value of C_2 is increased by 1. Assume $i = 2$ now. Then, the head of the input tape of T moves to the right. For every e_1 it sees, the value of C_1 is increased by 1. For every e_3 it sees, the value of C_2 is decreased by 1. This process stops when the head sees an e_2. It also stops when the value of C_2 is 0 and the head sees an e_3. In this case, the symbols that e_2 and e_3 stand for, are interchanged. Assume $i = 1$ now. Then, the value of C_1 is decreased by 1. If the value of C_1 is 0 now, the symbols that e_1 and e_2 stand for, are interchanged. Then, the head of the input tape of T moves to the right. For every e_1 it sees, beginning with the cell the move starts on, the value of C_1 is increased by 1. For every e_3 the head sees, the value of C_2 is decreased by 1. This process stops when the head sees an e_2. It also stops when the value of C_2 is 0 and the head sees an e_3. In this case, the symbols that e_2 and e_3 stand for, are interchanged. After that, the next step of the computation of A is simulated. This way, T simulates the computation of A on the input w in time $O(|w|)$ and space $O(\log(|w|))$. $\qquad\square$

For binary alphabets we have:

Proposition 8.30. *Every subset of $\{a,b\}^*$ that is in* **NTRANS** *is accepted by a counter automaton. Every subset of $\{a,b\}^*$ that is in the class* **DTRANS** *is accepted by a λ-free deterministic counter automaton.*

Proof. The first statement of the proposition is not hard to see. Also, we easily get that every subset of $\{a,b\}^*$ that is in **DTRANS** is accepted by a deterministic counter automaton. Let $A = (Q, \{a,b\}, \tau, \{s\}, F, \delta)$ be a DFAwtl such that for all $(q,c) \in Q \times \{a,b\}$ we have $\delta(q,c) = \emptyset$ if and only if $c \in \tau(q)$. To see how we get a λ-free deterministic counter automaton B with $L(B) = L(A)$, we only have to take a closer look at the following situation. Assume $\tau(s) = \{a\}$ and that there are $\ell, M \in \mathbb{N}$ such that for every $m \geq M$ there is a $p \in Q$ with $\left(s, a^m b^\ell\right) \vdash_A^* (p, \lambda)$. Set

$$\ell_0 = \min\left(\left\{\ell \in \mathbb{N} \mid \exists M \in \mathbb{N} : \forall m \geq M : \exists p \in Q : \left(s, a^m b^\ell\right) \vdash_A^* (p, \lambda)\right\}\right)$$

and

$$M_0 = \min\left(\left\{M \in \mathbb{N} \mid \exists p \in Q : \forall m \geq M : \left(s, a^m b^{\ell_0}\right) \vdash_A^* \left(p, a^{m-M}\right)\right\}\right).$$

Let $p \in Q$ such that $\left(s, a^{M_0} b^{\ell_0}\right) \vdash_A^* (p, \lambda)$. Set

$$i_0 = \min\left(\left\{i \in \mathbb{N} \mid \exists j \in \mathbb{N}_{>0} : \exists r \in Q : (p, a^{i+j}) \vdash_A^* \left(r, a^j\right) \vdash_A^* (r, \lambda)\right\}\right)$$

and

$$j_0 = \min\left(\left\{j \in \mathbb{N}_{>0} \mid \exists r \in Q : (p, a^{i+j}) \vdash_A^* \left(r, a^j\right) \vdash_A^* (r, \lambda)\right\}\right).$$

While reading an input of the form a^n, for an $n > 0$, the automaton B does never change its stack. If $n < M_0 + i_0$, the number n is remembered in the finite control of B. If, on the other hand, $n \geq M_0 + i_0$, the value $(n - (M_0 + i_0)) \bmod j_0$ is remembered in the finite control of B. Now, it should be clear how B can be constructed. □

Part IV

Outro

9 Conclusion and Directions for Further Research

In this final section we shortly summarize our findings on semilinear sets and variants of jumping finite automata and point out some directions for further research. We studied the descriptional complexity of operations on semilinear sets in the third chapter. Upper bounds for the parameters of the resulting semilinear set were given, when one of the operations intersection, complementation, or inverse homomorphism is applied to semilinear sets. This was done by a careful analysis of the proofs by Ginsburg and Spanier that semilinear sets are closed under the given operations [32]. Our bounds for intersection and inverse homomorphism are polynomial in the parameters of the operand sets and the max-norm of the homomorphism, while the bounds for complementation are double exponential in the parameters of the operand set. By additionally applying a normal form result for semilinear sets by To [73], we get single exponential bounds for complementation. Note that Chistikov and Haase independently also gave single exponential bounds for complementation [15]. However, they did not give any bound for the cardinality of the index set of the resulting semilinear set. As far as we know, no non-trivial lower bounds for the descriptional complexity of operations on semilinear sets are known; maybe this problem could be interesting for future research.

In the fourth chapter we introduced semirecognizable subsets of monoids as a generalization of recognizable sets and studied the semirecognizable subsets of the free commutative monoid $\mathbb{N}^k$, which are all finite unions of strongly semirecognizable linear subsets of $\mathbb{N}^k$. We characterized when a linear subset of $\mathbb{N}^k$ is semirecognizable in terms of rational cones. Lattices were introduced as special strongly semirecognizable subsets of $\mathbb{N}^k$ such that each strongly semirecognizable subset of $\mathbb{N}^k$ equals a lattice if only points for which all components are "large enough" are considered. We investigated Carathéodory-like decompositions of lattices. Namely, we gave a characterization in terms of rational cones when a lattice is a finite union of semirecognizable linear sets with linearly independent periods. This characterization was generalized to (strongly) semirecognizable subsets of $\mathbb{N}^k$. We showed that the geometric property in terms of rational cones that appears in our characterizations always holds in dimension at most three but does not always hold in higher dimensions. A characterization when a subset of $\mathbb{N}^k$ is a finite union of semirecognizable sets was given in terms of quasi-lattices. We studied the Parikh-preimages of semirecognizable subsets of $\mathbb{N}^k$. These are exactly the permutation closed semirecognizable languages, which form a subfamily of the permutation closed semilinear languages. We gave a new short proof using counter automata of the result by Latteux that each permutation closed semilinear language over a binary alphabet is context free [48]. Characterizations in terms of linear algebra when permutation closed semirecognizable languages are accepted by different types of pushdown automata were given. Already over a binary alphabet there are uncountable many strongly semirecognizable languages. Hence, for the future study of semirecognizable languages we suggest a restriction to special subclasses of this family. For example, the (deterministic) context-free semirecognizable languages could be investigated. Also, the regular or context-free languages with a (semi)recognizable Parikh-image could be objects of future research.

The operational state complexity of jumping finite automata was considered in the fifth chapter. Using the results from Chapter 3, we gave upper bounds for the number of states of a jumping finite automaton accepting the resulting language, when one of the operations intersection, complementation, or inverse homomorphism is applied to languages accepted by jumping finite automata. The bounds for intersection and inverse homomorphism are polynomial in the numbers of states of the operand automata and the absolute value of the homomorphism. The bound for complementation is single exponential in the number of states of the operand automaton. For the operational state complexity of jumping finite automata no non-trivial lower bounds are known. Another interesting open problem related to this topic is, if there exists an efficient algorithm for minimizing the number of states of a jumping finite automaton.

In the sixth chapter we showed that for every permutation closed language L and every $n \geq 0$ the language L is accepted by a right one-way jumping finite automaton with at most n accepting states if and only if L is n-semirecognizable. Closure properties of **ROWJ** were given and we proved characterizations of languages accepted by right one-way jumping finite automata when the languages are given as concatenations of two languages over disjoint alphabets or of a language and a word. Problems involving right one-way jumping finite automata were considered in the seventh chapter. While for right one-way jumping finite automata emptiness, finiteness, and universality are **NL**-complete, it is open if regularity, context-freeness, disjointness, inclusion, and equivalence are decidable. Also, it is not known whether there is an algorithm that minimizes the number of states of a right one-way jumping finite automaton. We showed that for a one-turn PDA A it is undecidable whether $L(A)$ is in **ROWJ** and whether $L(A)$ is in **pROWJ**. The problem whether a given word is the prefix (suffix, factor, sub-word, respectively) of a word accepted by a given right one-way jumping finite automaton, is in **PSPACE**. No non-trivial lower bounds for these problems are known. For a fixed input alphabet of cardinality at least two the problem to decide whether the language accepted by a given right one-way jumping finite automaton is permutation closed, is **NL**-complete. If the input alphabet is fixed, for right one-way jumping finite automata that accept permutation closed languages the word problem, emptiness, finiteness, universality, regularity, acceptance by different kinds of PDA s, disjointness, and inclusion are all in **L**. For a fixed input alphabet the number of accepting states of a given right one-way jumping finite automaton that accepts a permutation closed language can be minimized in logarithmic space. The total number of states of a right one-way jumping finite automaton that accepts a permutation closed language can be minimized in polynomial space. Even for a fixed input alphabet of cardinality at least two it is unknown whether the number of states of a right one-way jumping finite automaton that accepts a permutation closed language can be minimized in polynomial time.

Nondeterministic right one-way jumping finite automata were investigated in the eighth chapter. We proved that the Parikh-image of every language from **pMROWJ** is a quasi-lattice. Thus, a language over a binary alphabet is in **pMROWJ** if and only if it is a finite union of permutation closed semirecognizable languages. For an arbitrary alphabet Σ an $L \subseteq \Sigma^*$ is a finite union of permutation closed semirecognizable languages if and only if for all $w \in \Sigma^*$ it holds $L/^d w \in$ **pMROWJ**. It is unknown if **pMROWJ** is closed under the operation of disjoint quotient with a word. We showed that every permutation closed semilinear language is in **NROWJ** and that **MROWJ** and **NROWJ** are incomparable. Then, we introduced three different ways how λ-MNFA s can be interpreted as right one-way jumping finite automata. For all of these three interpretations λ-NFAs accept the same languages as λ-MNFAs. For two of these three interpretations λ-MNFAs can only accept the languages from **MNROWJ**. For the third interpretation more languages are accepted. Closure properties of the families **MROWJ**, **NROWJ**, **MNROWJ**, and λ_3**MNROWJ** have not yet been considered. Also, the decidability and complexity of problems involving nondeterministic right one-way jumping finite automata have not been studied yet. We compared right one-way jumping finite automata to finite-state acceptors with translucent letters and proved that the classes **DTRANS** \ λ_3**MNROWJ** and **ROWJ** \ **NTRANS** are not empty, that we have **NTRANS** $\subset$ **NST**$(\log(n), n^2/\log(n))$ and also **DTRANS** $\subset$ **DST**$(\log(n), n^2/\log(n))$, and that for all $p \in \mathbb{Q}$ with $0 < p \leq 1$ it holds **NTRANS** $\subset$ **NST**(n^p, n^{2-p}) and **DTRANS** $\subset$ **DST**(n^p, n^{2-p}). Conversely, there exists a language $L \in$ **DTRANS** such that for all functions $f, g : \mathbb{N} \to \mathbb{N}$ with $f(n) \cdot g(n) \in o(n^2)$ we get $L \notin$ **NST**$(f(n), g(n))$. It could be interesting to investigate finite-state acceptors with translucent letters with a slightly different acceptance condition: an input is only accepted when all letters are read at the end of the computation. It is not hard to see that in this mode the permutation closed languages accepted by deterministic finite-state acceptors with translucent letters coincide with **pROWJ**, the family of all permutation closed semirecognizable languages.

Bibliography

[1] Y. Bar-Hillel, M. Perles, and E. Shamir. On formal properties of simple phrase structure grammars. *Zeitschrift für Phonetik, Sprachwissenschaft und Kommunikationsforschung*, 14:143–177, 1961.

[2] S. Beier and M. Holzer. Decidability of right one-way jumping finite automata. In M. Hoshi and S. Seki, editors, *Proceedings of the 22nd International Conference on Developments in Language Theory*, number 11088 in LNCS, pages 109–120, Tokyo, Japan, September 2018. Springer.

[3] S. Beier and M. Holzer. Properties of right one-way jumping finite automata. In S. Konstantinidis and G. Pighizzini, editors, *Proceedings of the 20th International Workshop on Descriptional Complexity of Formal Systems*, number 10952 in LNCS, pages 11–23, Halifax, Nova Scotia, Canada, July 2018. Springer.

[4] S. Beier and M. Holzer. Nondeterministic right one-way jumping finite automata (extended abstract). In M. Hospodár, G. Jirásková, and S. Konstantinidis, editors, *Proceedings of the 21th International Workshop on Descriptional Complexity of Formal Systems*, number 11612 in LNCS, pages 74–85, Košice, Slovakia, July 2019. Springer.

[5] S. Beier and M. Holzer. Properties of right one-way jumping finite automata. *Theoretical Computer Science*, 798:78–94, 2019.

[6] S. Beier and M. Holzer. Semi-linear lattices and right one-way jumping finite automata. In M. Hospodár and G. Jirásková, editors, *Proceedings of the 24th Conference on Implemenation and Application of Automata*, number 11601 in LNCS, pages 70–82, Košice, Slovakia, July 2019. Springer.

[7] S. Beier, M. Holzer, and M. Kutrib. On the descriptional complexity of operations on semilinear sets. In E. Csuhaj-Varjú, P. Dömösi, and G. Vaszil, editors, *Proceedings of the 15th International Conference on Automata and Formal Languages*, number 252 in EPTCS, pages 41–55, Debrecen, Hungary, September 2017.

[8] S. Beier, M. Holzer, and M. Kutrib. Operational state complexity and decidability of jumping finite automata. In É. Charlier, J. Leroy, and M. Rigo, editors, *Proceedings of the 21st International Conference on Developments in Language Theory*, number 10396 in LNCS, pages 96–108, Liège, Belgium, August 2017. Springer.

[9] S. Beier, M. Holzer, and M. Kutrib. Operational state complexity and decidability of jumping finite automata. *International Journal of Foundations of Computer Science*, 30(1):5–27, 2019.

[10] S. Bensch, H. Bordihn, M. Holzer, and M. Kutrib. Deterministic input-reversal and input-revolving finite automata. In C. Martín-Vide, F. Otto, and H. Fernau, editors, *Proceedings of the 2nd International Conference Language and Automata Theory and Applications*, number 5196 in LNCS, pages 113–124, Tarragona, Spain, March 2008. Springer.

[11] S. Bensch, H. Bordihn, M. Holzer, and M. Kutrib. On input-revolving deterministic and nondeterministic finite automata. *Information and Computation*, 207(11):1140–1155, November 2009.

[12] H. Bordihn, M. Holzer, and M. Kutrib. Input reversals and iterated pushdown automata—a new characterization of Khabbaz geometric hierarchy of languages. In C. S. Calude, E. Calude, and M. J. Dinneen, editors, *Proceedings of the 8th International Conference on Developments in Language Theory*, number 3340 in LNCS, pages 102–113, Auckland, New Zealand, December 2004. Springer.

[13] H. Bordihn, M. Holzer, and M. Kutrib. Revolving-input finite automata. In C. De Felice and A. Restivo, editors, *Proceedings of the 9th International Conference on Developments in Language Theory*, number 3572 in LNCS, pages 168–179, Palermo, Italy, July 2005.

Springer.

[14] H. Chigahara, S. Fazekas, and A. Yamamura. One-way jumping finite automata. *International Journal of Foundations of Computer Science*, 27(3):391–405, 2016.

[15] D. Chistikov and Ch. Haase. The taming of the semi-linear set. In I. Chatzigiannakis, M. Mitzenmacher, Y. Rabani, and D. Sangiorgi, editors, *Proceedings of the 43rd International Colloquium on Automata, Languages, and Programming*, volume 55 of *Leibniz International Proceedings in Informatics*, pages 128:1–128:13, Rome, Italy, August 2016. Schloss Dagstuhl–Leibniz-Zentrum für Informatik, Dagstuhl, Germany.

[16] N. Chomsky. Three models for the description of language. *IRE Transactions on Information Theory*, 2(2):113–124, 1956.

[17] N. Chomsky. On certain formal properties of grammars. *Information and Control*, 2:137–167, June 1959.

[18] N. Chomsky. Context-free grammars and pushdown storage. Quarterly Progressive Report 65, Research Laboratory of Electronics, Massachusetts Institute of Technology, 1962.

[19] N. Chomsky and G. A. Miller. Finite state languages. *Information and Control*, 1(2):91–112, 1958.

[20] M. Chrobak. Errata to: Finite automata and unary languages. *Theoretical Computer Science*, 302:497–498, June 2003.

[21] J. H. Conway and N. J. A. Sloane. *Sphere Packings, Lattices and Groups*, volume 290 (3rd ed.) of *Grundlehren der Mathematischen Wissenschaften*. Springer, 1999.

[22] L. E. Dickson. Finiteness of the odd perfect and primitive abundant numbers with n distinct prime factors. *American Journal of Mathematics*, 35(4):413–422, October 1913.

[23] A. Ehrenfeucht, D. Haussler, and G. Rozenberg. On regularity of context-free languages. *Theoretical Computer Science*, 27:311–332,

1983.

[24] S. Eilenberg and M.-P. Schützenberger. Rational sets in commutative monoids. *Journal of Algebra*, 13(2):173–191, 1969.

[25] C. C. Elgot and J. E. Mezei. On relations defined by finite automata. *IBM Journal*, 10:47–68, 1965.

[26] R. J. Evey. Application of pushdown-store machines. In J. D. Tupac, editor, *AFIPS Conference Proceedings of the 1963 Fall Joint Computer Conference*, volume 24 of *Joint Computer Conference Proceedings*, pages 215–227, Las Vegas, USA, 1963. Spartan Books.

[27] S. Z. Fazekas, K. Hoshi, and A. Yamamura. Enhancement of automata with jumping modes. In A. Castillo-Ramirez and P. P. B. de Oliveira, editors, *Cellular Automata and Discrete Complex Systems – Proceedings of the 25th International Workshop on Cellular Automata and Discrete Complex Systems AUTOMATA 2019*, volume 11525 of *LNCS*, pages 62–76, Guadalajara, Mexico, 2019. Springer-Verlag.

[28] H. Fernau, M. Paramasivan, and M. L. Schmid. Jumping finite automata: Characterizations and complexity. In F. Drewes, editor, *Proceedings of the 20th Conference on Implementation and Application of Automata*, number 9223 in LNCS, pages 89–101, Umeå, Sweden, August 2015. Springer.

[29] H. Fernau, M. Paramasivan, M. L. Schmid, and V. Vorel. Characterization and complexity results on jumping finite automata. *Theoretical Computer Science*, 679:31–52, 2017.

[30] Y. Gao, N. Moreira, R. Reis, and S. Yu. A survey on operational state complexity. *Journal of Automata, Languages and Combinatorics*, 21(4):251–310, 2016.

[31] S. Ginsburg and S. Greibach. Deterministic context free languages. *Information and Control*, 9(6):620–648, December 1966.

[32] S. Ginsburg and E. H. Spanier. Bounded ALGOL-like languages. *Transactions of the American Matematical Society*, 113:333–368, 1964.

[33] S. Ginsburg and E. H. Spanier. Bounded regular sets. *Proceedings of the American Mathematical Society*, 17(5):1043–1049, October 1966.

[34] S. Ginsburg and E. H. Spanier. Semigroups, Presburger formulas, and languages. *Pacific Journal of Mathematics*, 16(2):285–296, 1966.

[35] J. Gubeladze and M. Michałek. The poset of rational cones. *Pacific Journal of Mathematics*, 292(1):103–115, January 2018.

[36] L. H. Haines. *Generation and Recognition of Formal Languages*. PhD Thesis. Massachusetts Institute of Technology, Department of Mathematics, Cambridge, Massachusetts, USA, 1965.

[37] M. A. Harrison. *Introduction to Formal Language Theory*. Addison-Wesley, 1978.

[38] D. Hilbert and P. Bernays. *Grundlagen der Mathematik*. Edward Brothers Inc., Ann Arbor, Michigan, USA, 1944.

[39] J. E. Hopcroft and J. D. Ullman. *Introduction to Automata Theory, Languages and Computation*. Addison-Wesley, 1979.

[40] T.-D. Huynh. The complexity of semilinear sets. In J. de Bakker and J. van Leeuwen, editors, *Proceedings of the 7th International Colloquim Automata, Languages and Programming*, number 85 in LNCS, pages 324–337, Boordwijkerhout, the Netherlands, July 1980. Springer.

[41] Y. Igarashi. A pumping lemma for real-time deterministic context-free languages. *Theoretical Computer Science*, 36:89–97, 1985.

[42] N. Immerman. Nondeterministic space is closed under complementation. *SIAM Journal on Computing*, 17(5):935–938, October 1988.

[43] R. Ito. Every semilinear set is a finite union of disjoint linear sets. *Journal of Computer and System Sciences*, 3(2):221–231, 1969.

[44] S. C. Kleene. Representation of events in nerve nets and finite automata. In C. E. Shannon and J. McCarthy, editors, *Automata studies*, volume 34 of *Annals of Mathematics Studies*, pages 2–42. Princeton University Press, 1956.

[45] E. Kopczyński. Complexity of problems of commutative grammars. *Logical Methods in Computer Science*, 11(1):Paper 9, March 2015.

[46] S.-Y. Kuroda. Classes of languages and linear bounded automata. *Information and Control*, 7(2):207–223, June 1964.

[47] R. Laing and J. B. Wright. Commutative machines. Technical report, University of Michigan, Ann Arbor, Michigan, December 1962.

[48] M. Latteux. Cônes rationnels commutatifs. *Journal of Computer and System Sciences*, 18(3):307–333, 1979.

[49] G. J. Lavado, G. Pighizzini, and S. Seki. Converting nondeterministic automata and context-free grammars into Parikh equivalent one-way and two-way deterministic automata. *Information and Computation*, 228–229:1–15, July 2013.

[50] G. J. Lavado, G. Pighizzini, and S. Seki. Operational state complexity under Parikh equivalence. In H. Jürgensen, J. Karhumäki, and A. Okhotin, editors, *Proceedings of the 16th International Workshop on Descriptional Complexity of Formal Systems*, number 8614 in LNCS, pages 294–305, Turku, Finland, August 2014. Springer.

[51] W. S. McCulloch and W. H. Pitts. A logical calculus of the ideas immanent in nervous activity. *Bulletin of Mathematical Biophysics*, 5:115–133, 1943.

[52] J. D. McKnight. Kleene quotient theorem. *Pacific Journal of Mathematics*, pages 1343–1352, 1964.

[53] A. Meduna and P. Zemek. Jumping finite automata. *International Journal of Foundations of Computer Science*, 23(7):1555–1578, 2012.

[54] A. Meduna and P. Zemek. *Regulated Grammars and Automata*, chapter 17: Jumping Finite Automata, pages 567–585. Springer, 2014.

[55] J. Myhill. Linearly bounded automata. Wadd Technical Note 60-165, Wright Patterson Air Force Base, Ohio, USA, June 1960.

[56] J. R. Myhill. *Finite automata and the representation of events.*

Technical report WADD TR-57-624. Wright Patterson Air Force Base, Ohio, USA, 1957.

[57] B. Nagy and L. Kovács. Finite automata with translucent letters applied in natural and formal language theory. In N.T. Nguyen, editor, *TCCI XVII 2014*, volume 8790 of *LNCS*, pages 107–127, Berlin Heidelberg, 2014. Springer-Verlag.

[58] B. Nagy and F. Otto. Finite-state acceptors with translucent letters. In G. Bel-Enguix, V. Dahl, and A. O. De La Puente, editors, *1st Workshop on AI Methods for Interdisciplinary Research in Language and Biology—BILC*, pages 3–13, Rome, Italy, January 2011. SciTePress.

[59] A. Nerode. Linear automaton transformations. *Proceedings of the American Mathematical Society*, 9(4):541–544, 1958.

[60] A. G. Oettinger. Automatic syntactic analysis and the pushdown store. In R. Jakobson, editor, *Structure of Language and its Mathematical Aspects*, volume 12 of *Proceedings of Symposia in Applied Mathematics*, pages 104–129, Providence, Rhode Island, USA, 1961. AMS.

[61] R. J. Parikh. Language-generating devices. *Quarterly Progress Report No. 60, Research Laboratory of Electronics, Massachusetts Institute of Technology*, pages 199–212, 1961.

[62] R. J. Parikh. On context-free languages. *Journal of the ACM*, 13(4):570–581, October 1966.

[63] W. J. Paul. *Komplexitätstheorie*. Teubner, 1978.

[64] M. Presburger. Über die Vollständigkeit eines gewissen Systems der Arithmetik ganzer Zahlen, in welchem die Addition als einzige Operation hervortritt. *Sprawozdanie z I Kongresu metematyków slowiańskich*, 395:92–101, 1930.

[65] M. O. Rabin and D. Scott. Finite automata and their decision problems. *IBM Journal of Research and Development*, 3:114–125, 1959.

[66] M. Rigo. The commutative closure of a binary slip-language is context-

free: a new proof. *Discrete Applied Mathematics*, 131:665–672, 2003.

[67] A. Salomaa. *Formal Languages*. ACM Monograph Series. Academic Press, 1973.

[68] W. J. Savitch. Relationships between nondeterministic and deterministic tape complexities. *Journal of Computer and System Sciences*, 4(2):177–192, April 1970.

[69] M.-P. Schützenberger. On context-free languages and push-down automata. *Information and Control*, 6(3):246–264, 1963.

[70] G. Sénizergues. $L(A) = L(B)$? Decidability results from complete formal systems. *Theoretical Computer Science*, 251(1–2):1–166, January 2001.

[71] M. Studený. Convex cones in finite-dimensional real vector spaces. *Kybernetika*, 29(2):180–200, 1993.

[72] R. Szelepcsényi. The method of forced enumeration for nondeterministic automata. *Acta Informatica*, 26(3):279–284, November 1988.

[73] A. W. To. Parikh images of regular languages: Complexity and applications. http://arxiv.org/abs/1002.1464v2, LFCS, School of Informatics, University of Edinburgh, February 2010.

[74] A. M. Turing. On computable numbers, with an application to the Entscheidungsproblem. *Proceedings of the London Mathematical Society*, 42(1):230–265, 1936.

[75] J. von zur Gathen and M. Sieveking. A bound on solutions of linear integer equalities and inequalities. *Proceedings of the American Mathematical Society*, 72(1):155–158, October 1978.

[76] S. Yu. A pumping lemma for deterministic context-free languages. *Information Processing Letters*, 31:47–51, 1989.

Index